WATER RESOURCE PLANNING, DEVELOPMENT AND MANAGEMENT SERIES

WATER INFRASTRUCTURE

WATER RESOURCE PLANNING, DEVELOPMENT AND MANAGEMENT SERIES

Planning and Managing of Water Resources
Peter A. Unwal (Editor)
2006. ISBN: 1-59454-757-2

Water Resources Research Progress
Liam N. Robinson (Editor)
2008. ISBN: 1-60021-973-x

Trends in Water Resources Research
Alan B. Prescott and Thomas U. Barkely (Editors)
2008. ISBN: 978-1-60456-038-1

Integrated Water Resource Management in the Kurdistan Region
Almas Heshmati
2009. ISBN: 978-1-60741-295-3

Safeguarding the Nation's Drinking Water
Juan A. Schrock (Editor)
2010. ISBN: 978-1-60741-234-2

Water Infrastructure
Harald Nystroem (Editor)
2010. ISBN: 978-1-60692-479-2

WATER RESOURCE PLANNING, DEVELOPMENT AND MANAGEMENT SERIES

WATER INFRASTRUCTURE

HARALD NYSTROEM
EDITOR

Nova Science Publishers, Inc.
New York

For permission to use material from this book please contact us:
Telephone 631-231-7269; Fax 631-231-8175
Web Site: http://www.novapublishers.com

LIBRARY OF CONGRESS CATALOGING-IN-PUBLICATION DATA

Available upon request

ISBN: 978-1-60692-479-2

Published by Nova Science Publishers, Inc. + New York

CONTENTS

Preface		**vii**
Chapter 1	Aging Infrastructure: Dam Safety *Nic Lane*	**1**
Chapter 2	The Bureau of Reclamation's Aging Infrastructure *Nic Lane*	**23**
Chapter 3	Senate Committee on Energy and Natural Resources. Water and Power Subcommittee Hearing: Testimony on the Increasing Number of Issues Associated with Aging Water Resource Infrastructure That Are Operated and Maintained, or Owned, by the United States Bureau of Reclamation (SD-366) *Wesley Randal "Randy" Reed*	**37**
Chapter 4	Testimony of Tony Willardson, Western States Water Council on Behalf of the Western Governors' Association and the Western States Water Council before the Subcommittee on Water and Power of the Energy and Natural Resources Committee United States Senate *Tony Willardson*	**51**

Chapter 5 Testimony before the United States Senate Energy and Natural Resources Committee's, Subcommittee on Water and Power, An Oversight Hearing on Aging Infrastructure **59**
Thomas F. Donnelly

Chapter 6 Statement of Robert W. Johnson, Bureau of Reclamation, U.S. Department of the Interior, before the Committee on Energy and Natural Resources, Subcommittee on Water and Power U.S. Senate, Reclamation's Aging Infrastructure **65**
Robert W. Johnson

Chapter 7 Testimony of Mayor Todd Cutler, City of Fernley, Nevada, Senate Subcommittee on Water and Power **71**
Todd Cutler

Chapter 8 Water Infrastructure Financing: History of EPA Appropriations **75**
Claudia Copeland

Chapter 9 Statement of Charles I. McGinnis before Committee on Organizing to Manage Construction and Infrastructure in the 21st Century, Bureau of Reclamation Board on Infrastructure and the Constructed Environment Division on Engineering and Physical Sciences, The National Academies **117**
Charles I. McGinnis

Chapter Sources **129**

Index **131**

PREFACE

The Bureau of Reclamation is responsible for the construction of most of the large irrigation and water resources infrastructure in the West. Reclamation manages water resource facilities in 17 western states with an original development cost of over $20.0 billion. Furthermore, Reclamation is is over 100 years old. This aging infrastructure requires increased maintenance and replacement efforts and expenditures. This book describes Reclamation's approach to managing aging infrastructure. There is also a discussion of four specific approaches to managing Reclamation's aging infrastructure through legislative action. Furthermore, the principal federal program to aid municipal wastewater treatment plant construction is authorized in the Clean Water Act (CWA). This book explores the funding for EPA wastewater assistance and some of the funded projects which are not authorized in the Clean Water Act nor the Safe Drinking Water Act. This book also examines the urgent actions needed to be taken for additional funding to resolve any deficiencies in dams. While dams have multiple benefits, they can also present a risk to public safety and economic infrastructure. This book consists of public documents which have been located, gathered, combined, reformatted, and enhanced with a subject index, selectively edited and bound to provide easy access.

Chapter 1 - While dams have multiple benefits (balanced against financial and environmental costs), they can also present a risk to public safety and economic infrastructure. This risk stems from two sources: the possibility of a dam failure and the damage it would cause. Although dam failures are infrequent, age, construction deficiencies, inadequate maintenance, and seismic or weather events contribute to the possibility. To reduce the risk, regular inspections are necessary to identify potential problems. Corrective action can then be taken to remedy those deficiencies. Congress is often called upon to fund remedial actions, as a way to prevent the larger catastrophes. The 110th Congress will likelysee proposals for improving dam safety and may oversee existing safety programs.

To identify deficiencies that could cause dam failures, the federal government established inspection requirements for the nation's federal dams. Once deficiencies are identified, mostagencies finance repairs through their operation and maintenance accounts. Funding mechanisms vary for larger rehabilitation activities. At the Bureau of Reclamation, for example, most larger repairs are conducted with annual appropriations to its dam safety program. At some other agencies, dam rehabilitation must compete with other construction projects for funding.

The federal agencies with dam safety responsibilities include the Tennessee Valley Authority, the Federal Energy Regulatory Commission (FERC), and the Departments of Agriculture, Defense, Energy, the Interior, Labor, and State. At nonfederal dams, safety is generally a state responsibility, though some federal assistance has been provided. The National Dam Safety Program, which is authorized through FY2011 by P.L. 109-460, helps states improve their dam safety programs and train inspectors. In addition, FERC and the Department of Labor's Mine Safety and Health Administration require regular inspections at the nonfederal dams within their jurisdiction. Even so, there are concerns that most state dam safety programs have inadequate staff and funds to effectively inspect or monitor all of the dams for which they are responsible. Further, there are concerns that states, local governments, and other nonfederal dam owners may not have the financial resources to maintain and rehabilitate their dams.

The collapse of the I-35W bridge in Minnesota highlighted the potential for unexpected infrastructure failure. This may result in a review of the safety of other elements of our nation's inventory of critical infrastructure such as dams, levees, tunnels, and bridges — and a call for additional funding to resolve any deficiencies. Congress has periodically been urged to provide federal support for rehabilitation work at nonfederal dams. Demand for such assistance is likely to increase, but currently no federal policy describes the conditions under which federal funding is appropriate, nor has Congress established criteria for prioritizing funding among nonfederal projects.

Chapter 2 - The Bureau of Reclamation (Reclamation) is responsible for the construction of most of the large irrigation and water resources infrastructure in the West. Reclamation manages water resource facilities in 17 western states with an original development cost of over $20.0 billion. Reclamation is over 100 years old, and many of its facilities now have an average age of over 50 years. This aging infrastructure requires increased maintenance and replacement efforts and expenditures.

Reclamation appears to have a well documented plan to assess the management needs of its portfolio of aging infrastructure. However, water resources infrastructure management objectives require prioritization due in part to a finite budget. This inevitably leads to conflicts over project funding priorities. As Reclamation's portfolio of infrastructure continues to age, these conflicts are likely to arise more often. Further, in the case of structures whose operations and maintenance responsibilities have been transferred to the local project beneficiaries, the increased maintenance and replacement efforts and expenditures associated with aging infrastructure — such as more expensive recapitalization projects — could motivate the beneficiaries to seek federal funds and congressional support to address project funding issues.

This report describes Reclamation's approach to managing aging infrastructure as well as that of two other agencies — the Army Corps of Engineers and the Natural Resources Conservation Service — involved with significant portfolios of dams and related infrastructure. Additionally, there is discussion of four specific approaches to managing Reclamation's aging infrastructure through legislative action.

Chapter 3 – This is a testimony before the Senate Committee on Energy and Natural Resources.

Chapter 4 – This is a testimony of Tony Willardson, Western States Water Council before the Senate Committee on Energy and Natural Resources.

Chapter 5 - This is a testimony of Thomas F. Donnelly, National Water Resources Association before the Senate Committee on Energy and Natural Resources.

Chapter 6 - This is a testimony of Robert W. Johnson, Bureau of Reclamation, United States Department of the Interior before the Senate Committee on Energy and Natural Resources.

Chapter 7 - This is a testimony of Todd Cutler, City of Fernley, Nevada, before the Senate Committee on Energy and Natural Resources.

Chapter 8 - The principal federal program to aid municipal wastewater treatment plant construction is authorized in the Clean Water Act (CWA). Established as a grant program in 1972, it now capitalizes state loan programs. Authorizations since 1972 have totaled $65 billion, while appropriations have totaled $78.3 billion. It has represented 25-30% of total funds appropriated to the Environmental Protection Agency (EPA) in recent years.

In appropriations legislation, funding for EPA wastewater assistance is contained in the measure providing funds for the Department of the Interior, Environment, and Related Agencies, which includes EPA. Within the portion of that bill which funds EPA, wastewater treatment assistance is specified in an account now called State and Tribal Assistance Grants (STAG). Three trends in the funding of this account are most prominent: inclusion of non-infrastructure environmental grants to states, beginning in FY1993; increasing number and amount of special purpose grants since FY1989; and the addition of grant assistance for drinking water treatment projects in FY1997. This report summarizes, in chronological order, congressional activity to fund items in this account since 1987.

Prior to the 1987 amendments, wastewater treatment assistance was provided in the form of grants made to municipalities. The federal share of project costs was generally 55%; state and local governments were responsible for the remaining 45%. The 1987 amendments altered this arrangement by replacing the traditional grant program with one that provides federal grants to capitalize state clean water loan programs, or state revolving funds (SRFs). As a general matter, states and cities support the program changes and the shift to a loan program that was intended to provide long-term funding for water quality and wastewater construction activities. However, the change means that local communities now are responsible for 100% of projects costs, rather than 45%, because they are required to repay loans to states. The greater financial burden of the act's loan program on some cities has caused some to seek continued grant funding.

This has been particularly evident in the appropriations process where, in recent years, Congress has reserved as much as 30% of funds in the STAG account for special purpose grants directed to specified communities. Since FY2000, appropriators have awarded earmarks to a larger total number of projects, resulting in more communities receiving such grants, but at the same time receiving smaller amounts of funds, on average. Most of the funded projects are not authorized in the Clean Water Act or the Safe Drinking Water Act. State water quality officials, state infrastructure financing officials, and EPA have objected to this practice, since it reduces the amount of funding for state SRF programs. Since FY1997, the STAG account also has been used to fund a drinking water SRF grant program established by Congress in 1996. Appropriations for the drinking water SRF program through FY2008 have totaled $10.3 billion.

Chapter 9 – This is a testimony of Charles I. McGinnis, United States Army Corps of Engineers before the Subcommittee on Water and Power Committee on Energy and Natural Resources, United States Senate.

In: Water Infrastructure
Editor: Harald Nystroem
ISBN: 978-1-60692-479-2

Chapter 1

AGING INFRASTRUCTURE: DAM SAFETY*

Nic Lane

SUMMARY

While dams have multiple benefits (balanced against financial and environmental costs), they can also present a risk to public safety and economic infrastructure. This risk stems from two sources: the possibility of a dam failure and the damage it would cause. Although dam failures are infrequent, age, construction deficiencies, inadequate maintenance, and seismic or weather events contribute to the possibility. To reduce the risk, regular inspections are necessary to identify potential problems. Corrective action can then be taken to remedy those deficiencies. Congress is often called upon to fund remedial actions, as a way to prevent the larger catastrophes. The 110th Congress will likelysee proposals for improving dam safety and may oversee existing safety programs.

To identify deficiencies that could cause dam failures, the federal government established inspection requirements for the nation's federal dams. Once deficiencies are identified, mostagencies finance repairs through their operation and maintenance accounts. Funding mechanisms vary for larger rehabilitation activities. At the Bureau of Reclamation, for example, most larger repairs are conducted with annual appropriations to its dam safety

---

* This is an edited, reformatted and augmented version of a CRS Report for Congress publication dated March 2008.

program. At some other agencies, dam rehabilitation must compete with other construction projects for funding.

The federal agencies with dam safety responsibilities include the Tennessee Valley Authority, the Federal Energy Regulatory Commission (FERC), and the Departments of Agriculture, Defense, Energy, the Interior, Labor, and State. At nonfederal dams, safety is generally a state responsibility, though some federal assistance has been provided. The National Dam Safety Program, which is authorized through FY2011 by P.L. 109-460, helps states improve their dam safety programs and train inspectors. In addition, FERC and the Department of Labor's Mine Safety and Health Administration require regular inspections at the nonfederal dams within their jurisdiction. Even so, there are concerns that most state dam safety programs have inadequate staff and funds to effectively inspect or monitor all of the dams for which they are responsible. Further, there are concerns that states, local governments, and other nonfederal dam owners may not have the financial resources to maintain and rehabilitate their dams.

The collapse of the I-35W bridge in Minnesota highlighted the potential for unexpected infrastructure failure. This may result in a review of the safety of other elements of our nation's inventory of critical infrastructure such as dams, levees, tunnels, and bridges — and a call for additional funding to resolve any deficiencies. Congress has periodically been urged to provide federal support for rehabilitation work at nonfederal dams. Demand for such assistance is likely to increase, but currently no federal policy describes the conditions under which federal funding is appropriate, nor has Congress established criteria for prioritizing funding among nonfederal projects.

INTRODUCTION[1]

Dams provide broad economic and social benefits, including flood control, hydroelectric power, recreation, navigation, and water supply for drinking, irrigation, and industrial uses. Dams also entail financial costs (for construction and for operation and maintenance) and sometimes environmental costs (e.g., loss of riverine habitat). Dams also can present a risk to public safety, local and regional economies, and the environment.

Prior to September 11, 2001, Congress had expressed an increasing interest over several decades in dam safety. In recent years, congressional interest has focused largely on securing and protecting U.S. dams and water storage facilities from terror attacks. Hurricane Katrina and the disastrous failure of levees that once protected New Orleans, however, have renewed congressional interest in the structural integrity of major water infrastructure.

The modern period of congressional concern began in the 1970s with dam failures that resulted in loss of life and billions of dollars in property losses.[2] Congress and private groups interested in dam safety noted that, while states and localities are responsible for the maintenance and safetyof 95% of the nation's dams, large numbers of older dams lacked the maintenance needed to guarantee operational integrity and prevent failure. These aging dams represented then — and continue to present — a potential hazard to downstream populations.

A first essential task was to develop accurate data on the nation's dams: their number, type, structural condition, and other information useful for making decisions about dam safety policy and priorities. This was achieved in 1972 with the creation of the National Inventory of Dams (NID)[3] — a computerized, periodically updated[4] catalogue of U.S. dams, maintained by the U.S. Army Corps of Engineers, that presently lists over 79,000 public and private dams.[5] For the purposes of the NID, a dam is defined as "any artificial barrier that has the ability to impound water ... for the purpose of storage or control of water" that (1) is greater than 25 feet in height with a storage capacity of more than 15 acre-feet[6] (af), (2) is greater than 6 feet in height with a storage capacity of more than 50 af, or (3) poses a significant threat to human life or property should it fail.

Federal and nonfederal dams in the NID were constructed for a variety of purposes. By far, the greatest number of dams — more than 40% — were constructed primarily for recreational purposes. Other purposes served by dams, in descending order, include fire protection (20.0%), flood control (19.8%); irrigation (11.8%); water supply (9.3%); fish and wildlife (4.2%); hydroelectric (2.8%); debris control (2.0%); mining (tailings dams; 1.8%); and navigation (0.1%). Many dams serve multiple purposes. For approximately 9.7% of the dams in the NID, the purposes are unspecified.[7]

Over 56% of dams in the NID are privately owned. (See **Figure 1**.) Slightly more than 20% are owned and operated by local governments. About 4.8 % of dams are owned by states and 2.4% by public utility companies. The federal government owns only 4.7% of all NID dams, but this small number (3,771) includes the dams many Americans view as iconic:the greathydroelectric dams of the West, like Grand Coulee and Hoover.[8] The ownership of some NID dams is not indicated in the database because that information was not reported to the Corps.

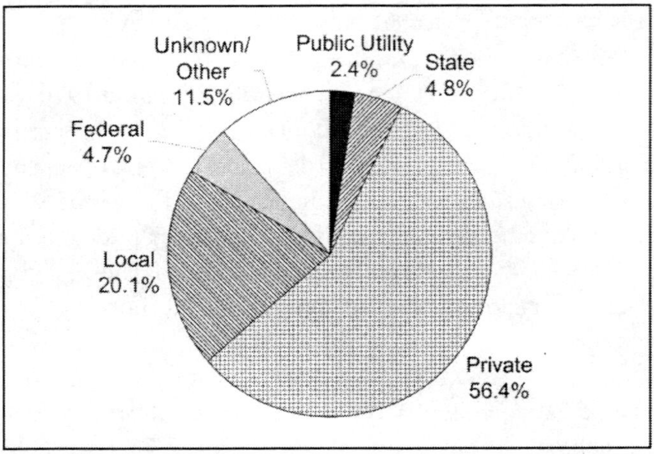

Figure 1. Dam Ownership

While the federal government owns less than 5% of NID dams, more than 30% of all dams in the NID inventory were funded, designed, or constructed with federal resources, most of them through the Department of Agriculture's Natural Resources Conservation Service (NRCS). NRCS's involvement in dam construction stems primarily from the Watershed Protection and Flood Prevention Act of 1954,[9] which authorized it to cooperate with states and local agencies to undertake works of improvement for flood prevention and other purposes. Under this act and an earlier law to build projects in 11 designated watersheds, NRCS has helped build more than 10,000 upstream flood control dams beginning in 1948. These are generally relatively small dams owned by public or private entities other than the federal government. These nonfederal entities are principally responsible for the dams' operation, maintenance, and security.[10]

Excluding the NRCS, which does not administer any dams, nine federal agencies operate, or regulate dam safety at approximately 8,500 sites (including non-NID dams).[11] Six federal agencies operate NID dams: the Department of Defense (862), Department of the Interior (751), Department of Agriculture (326), Tennessee Valley Authority (83), Department of Energy (16), and State Department (7). In addition, three agencies regulate but do not operate NID dams: FERC regulates 1,775 hydropower dams; the Department of Labor's Mine Safety and Health Administration regulates 745 dams; and the Nuclear Regulatory Commission regulates 11 dams. While dams administered by one agency are not generally regulated by another agency, there are cases where private hydroelectric projects, regulated by FERC, are located at federal dams.

Congressional interest in dam safety generally falls into three areas: (1) dam security and the potential for acts of terrorism at major U.S. dam sites; (2) prevention of potential dam failures due to structural deficiencies; and (3) recovery from dam failures. This report focuses on the second category, because it is a topic that will likely become more important as the nation's dams age; further, it has gained additional attention in the aftermath of Hurricane Katrina.

DAM FAILURE RISK

While dams have multiple benefits, their failure or misoperation could threaten public safety, local and regional economies, and the environment. Risk has two primary components: (1) the damage and deaths associated with dam failure or misoperation, and (2) the likelihood of such a failure.

Potential Damage

To quantify the potential harm associated with a dam's failure, the Interagency Committee on Dam Safety prepared a hazard potential classification system.[12] As described in **Table 1**, the three hazard ratings (low, significant, and high) do not indicate the likelihood of failure, but reflect the *amount and type of damage* that a failure would cause. Hazard ratings for each dam are included in the NID. From 2000 to 2006 the number of high-hazard dams increased from 9,921 to 11,811. According to the Federal Emergency Management Agency (FEMA), development below dams is the primary factor increasing dams' hazard potential.[13]

Likelihood of Failure

While catastrophic dam failures are fairly infrequent, states reported 1,090 dam safety incidents — including 125 failures — between 1999 and 2004.[14] A number of factors, including age, construction deficiencies, inadequate maintenance, and seismic or weather events, contribute to the likelihood of dam failure. For example, some failures are the direct result of flows larger than the dams were built to withstand. With the exception of seismic or weather events, age is a leading indicator of dam failure. In particular, the structural integrity and operational effectiveness of dams may deteriorate with age, and some older dams may not comply with current dam safety standards established in the 1970s.[15] Overall, more than 30% of all dams in the NID are at least 50 years old, the designed lifespan of many dams, and more than 17,000 will cross this threshold over the next 10 years. (See **Figure 2.**) According to the Association of State Dam Safety Officials, in 2003, approximately 3,243 U.S. dams had deficiencies that left them more susceptible to failure.[16] In 2000, another report estimated that more than $30 billion will be needed to repair and rehabilitate the nation's aging dams.[17]

MANAGING DAM SAFETY

Following dam failures at Buffalo Creek (WV, 1972), Teton Dam (ID, 1976), and Kelly Barnes Dam (GA, 1977), legislative and executive actions established a program for monitoring the nation's dams and set guidelinesfor dam safety at federal facilities. Subsequent legislation promotes state dam safety programs and facilitates rehabilitation activities at federal and some nonfederal dams.

Table 1. Hazard Level: Description and Number

Hazard Classification	Result of Failure or Misoperation	Number of Dams
High Hazard	— Loss of life is probable. — Other economic or environmental loss possible, but not necessary for this classification.	11,811
Significant Hazard	— No probable loss of human life. — Could result in economic loss, environmental damage, and disruption of lifeline facilities, etc.	13,407
Low Hazard	— No probable loss of human life. — Few economic or environmental losses; losses are generally limited to the owner.	54,349

Source: *NID Data*. The hazard level of 210 dams was not reported to the Corps for inclusion in the NID.

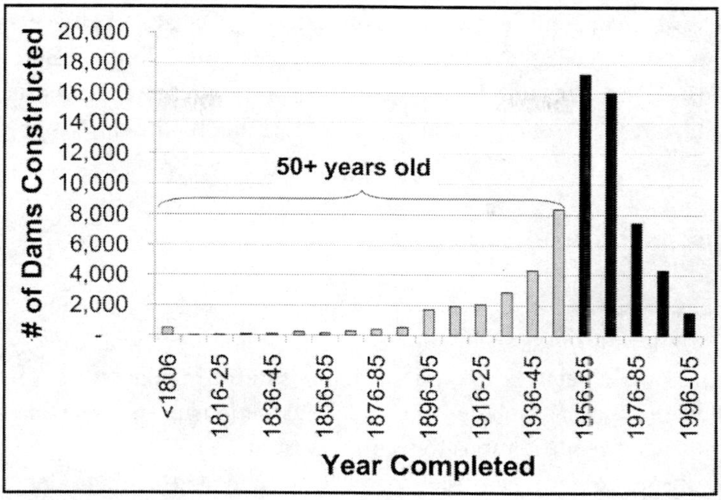

Figure 2. Construction of Federal and Nonfederal Dams

Federal Management

Through legislative and executive actions, the federal government has become involved in multiple areas of dam safety. First, in 1972, Congress passed the National Dam Inspection Act (P.L. 92-367), which authorized the Corps to undertake a program of national dam inspections and to establish the National Inventory of Dams. Activities under this law provided the impetus for broad scale monitoring and a centralized location for information on many of the nation's dams.

In 1977, following the failure of Teton Dam and Kelly Barnes Dam, President Jimmy Carter ordered a review of federal dam safety activities, and subsequently the *ad hoc* Interagency Committee on Dam Safety released safety guidelines for dams regulated by federal agencies.[18] The *Federal Guidelines for Dam Safety* establish a basic structure for agencies' dam safety programs. In general, the guidelines encourage federal agencies and dam owners regulated by federal agencies to abide by uniform safety standards. They direct that each agency with dam safety responsibilities have a dam safety officer and that the agencies coordinate with other agencies. They also establish guidelines for staff training, periodic evaluations, documenting dam safety activities, and operation and maintenance.

Congress also enacted the Reclamation Safety of Dams Act (P.L. 95-578, 43 U.S.C. §508) in 1978 to set more detailed guidelines for the Bureau of Reclamation.[19] This act authorizes Reclamation to preserve the structural stability of its dams and related facilities by performing modifications.

State Dam Safety Programs

States have primary responsibility for the safety of 95% of the nation's dams. According to the Association of State Dam Safety Officials, every state but Alabama and Delaware[20] has a dam safety regulatory program.[21] Typically these programs include "(1) safety evaluations of existing dams, (2) review of plans and specifications for dam construction and major repair work, (3) periodic inspections of construction work on new and existing dams, and (4) review and approval of emergency action plans."[22] Many state dam safety programs are poorly funded. In 2004, state budgets for dam safety averaged $742,000,[23] ranging from less than $50 per state-regulated dam (IA, IN, KS, MS, OK and TX) to more than $16,000 per dam regulated by Puerto Rico.[24]

Federal Support for State Dam Safety Programs

While federal activities in the 1970s generally focused on increasing dam safety information and strengthening requirements at the nation's federal dams, subsequent legislation began to address the safety of nonfederal dams. In 1996, Congress created the National Dam Safety Program (NDSP)[25] and assigned responsibility for administering it to FEMA. The NDSP is the nation's principal dam safety program; previously, there was no comprehensive national effort devoted to nonfederal dam safety and the safety of downstream populations.

Management of the NDSP

The NDSP is a mechanism for federal and state cooperation that includes an Interagency Committee on Dam Safety (ICODS) with the Director of FEMA as its chair. ICODS, which was initially formed in 1980, is composed of representatives from FERC, the International Boundary Water Commission, the Nuclear Regulatory Commission, the Tennessee Valley Authority, and the Departments of Agriculture, Defense, Energy, the Interior, Labor, and Homeland Security (FEMA). ICODS is responsible for coordinating information exchange among federal dam safety agencies.[26]

The act also established a National Dam Safety Review Board (NDSRB) consisting of five representatives appointed from federal agencies, five state dam safety officials, and one representative from the U.S. Society on Dams. All the representatives are appointed by the director of FEMA. This board advises the FEMA director on dam safety issues, including the allocation of grants.[27]

Assistance to States

The National Dam Safety Act establishes two state assistance programs with budget authorizations. These include the following:

- Training for State Inspectors. At the request of states, FEMA provides technical training to dam safety inspectors.
- Assistance to State Dam Safety Programs. States working toward or meeting minimal requirements are eligible for assistance grants. The objective of these grants is to upgrade state programs using the *ASDSO Model State Dam Safety Program* as a guide. The model state dam safety program includes a recommendation that dams be inspected at least every five years.

Allocation of state assistance grants is determined by the NDSRB and the director of FEMA. In FY2000 and FY2001, FEMA distributed $8 million to states with existing dam safety programs to assist with buying equipment, conducting dam inspections, and developing Emergency Action Plans (EAPs). This money is not available for rehabilitation activities.

Reauthorization of the NDSP

The NDSP was reauthorized in 2002 as the Dam Safety and Security Act of 2003 (P.L. 107-310, 43 U.S.C. §467). P.L. 107-310 made several changes to the existing program, including the addition of (1) a new goal for the program that encompasses dam security; (2) a goal for states to obtain authority to require owners to improve security; (3) a limitation on the authority of ICODS to exchange information among federal entities only; (4) a clearer definition of the role of the NDSRB in relation to the states, and encouragement to maintain an effective national program to enhance dam safety and protect human life and property; and (5) the addition of two new research components on (a) information technology to store, query, and distribute dam safety data related to dam performance (failures, large storm events, earthquakes, etc.), and (b) dam safety vulnerability assessments and management of sensitive dam information.

P.L. 107-310 authorized the program for an additional four years and increased the total authorized funding to $8.6 million annually through FY2006 with funds available until expended. Specified funding authorizations include an increase for research from $1.0 million to $1.5 million for each fiscal year. The program retains the authorized levels for dam safety training at $500,000, and increases authorized funding for staff from $400,000 to $600,000 for the same time period.

The enactment of P.L. 109-460 authorized an average of $9.96 million annually for FY2007 through FY2011 for FEMA-administered research, dam safety training, and maintaining the NID.

REDUCING DAM FAILURES

As the nation's dams age, they are likely to develop various deficiencies. For example, dams' foundations can show signs of seepage, cracking, and movement. To prevent failure or misoperation, these deficiencies must be identified and corrected.

Identifying Unsafe Dams

The first step toward rectifying dam-safety issues is to identify safety deficiencies. Such deficiencies are often identified by engineers during informal inspections, or during formal inspections conducted by senior engineers. Formal inspections are generally conducted after major seismic or weather events, and on a periodic basis. As described below, the frequency and type of periodic inspections varies across agencies.

Inspections of Federal Dams

Each of the six federalagencies thatoperate federal dams is responsible for maintaining dam safety by performing maintenance, inspections, and rehabilitation work. The *Federal Guidelines for Dam Safety*, established in 1979 by President Carter, provide basic guidance for agencies' dam safety programs. Specifically, this document recommends that agencies formally inspect each dam at least once every five years. As shown in **Table 2**, however, most agencies under the Department of the Interior require more frequent inspections. These inspections are typically funded through the agencies' operations and maintenance (O&M) budgets.

Inspection of Nonfederal Dams

Inspecting nonfederal dams is generally a state responsibility, but the states are often poorly funded. According to the Association of State Dam Safety Officials, 10 state regulators are needed for every 250 dams to do the job of carrying out their responsibilities.[28] However, the average number of dams per FTE is 395 and only two states (CA and FL) and one territory (PR) have the recommended number of staff.[29]

Table 2. Agency-Owned Dams by Hazard Level and Frequency of Formal Inspections

Agency	Number of Dams by Hazard Level			Frequency of Formal Inspections[a]
	Total	Significant	High	
Department of Agriculture				
Forest Service	326	70	29	5 years
Department of Defense				
Air Force	29	0	0	5 years
Army	191	24	30	5 years
Corps of Engineers	608	90	470	5 years
Navy	34	4	4	5 years
Department of the Interior				
Bureau of Reclamation	435	24	343	3 years
Bureau of Indian Affairs	64	19	44	3 years
Bureau of Land Management	2	2	0	3 years
Fish and Wildlife Service	177	22	11	2.5 years
National Park Service	72	30	13	5 years
Department of State				
International Boundary Water Commission	7	1	5	5 years
Tennessee Valley Authority	83	15	63	5 years
Department of Energy	16	2	1	5 years

Source: *NID Data.*

a. Inspection frequency obtained from agency officials in Nov. 2004.

Federal Involvement

While regulating nonfederal dams is generally a state responsibility, the federal government has assumed regulatory authority over certain nonfederal dams. As described below, two federal agencies — FERC and the Department of Labor's Mine Safety and Health Administration (MSHA) — participate in inspections of certain private dams. In addition, the NRCS may become involved with inspections at the dams it constructed.

Private Hydropower Dams. Under the 1920 Federal Power Act, FERC (formerly the Federal Power Commission) regulates more than 2,500 nonfederal hydropower dams.[30] Pursuant to the *Federal Guidelines on Dam Safety* and FERC Order 122,[31] FERC's regional engineers are to inspect each high-hazard dam annually and outside consultants are to inspect these dams every five years. Low-hazard dams are to be inspected every three years. The federal government recovers the costs of these and FERC's other activities from the hydropower industry. In general, FERC's dam safety program has received positive recognition.[32]

Private Mining Dams. Under the Mine Safety and Health Act,[33] MSHA regulates 745 private dams. Under these regulations, dams used for surface mining are normally inspected every two years and those used for underground coal mines are to be inspected every four years.[34]

NRCS-Constructed Flood Control Dams. Since 1948, the NRCS has constructed more than 10,500 flood control dams. These dams were turned over to local entities under contracts that stipulate the nonfederal responsibility for operation and maintenance of the dams. Therefore, the dam owners are responsible for conducting inspections pursuant to state regulations. As resources permit, however, NRCS may use funding appropriated under the Small Watershed Rehabilitation Act[35] to help dam owners assess specific structural or operational problems. In certain situations, NRCS will conduct an inspection as part of this assessment process.

As amended by §2505 of the 2002 farm bill (P.L. 107-171),[36] the Small Watershed Rehabilitation Program is authorized to receive both mandatory funding through the Commodity Credit Corporation and discretionary funding to be authorized through agriculture appropriations each year. Mandatory funding amounts were to start at $45 million in FY2003 and increase by $5 million each year through FY2007, while discretionary funding was authorized at $45 million in FY2003 and was to grow by $10 million each year through

FY2007. To date, congressional appropriators have prohibited mandatory funding each year while providing a portion of the discretionary funding that had been authorized. Actual appropriations were $29.8 million in FY2003, $29.6 million in FY2004, $27.5 million in FY2005, and $31.5 million in FY2006, $31.3 million in FY2007,[37] and $19.9 million for FY2008. The FY2009 budget estimate for this program is $5.9 million.[38]

DAM REHABILITATION

Federal Dam Rehabilitation

After dam safety deficiencies have been identified, rehabilitation activities should be undertaken. However, most federal agencies do not have funding available to immediately undertake all non-urgent repairs. Rather, they generally prioritize their rehabilitation needs — based on various forms of risk assessment — and schedule these activities in conjunction with the budget process. At some agencies, dam rehabilitation needs must compete for funding with other construction projects. Rehabilitation activities at the two major dam-owning agencies aredescribed below.

Department of Defense, Army Corps of Engineers

At the Corps, most dam deficiencies are addressed through the normal O&M procedures. However, "rehabilitation or modification of Corps' dams for safety purposes is accomplished through the Major Rehabilitation Program and the Dam Safety Assurance Program."[39] The purpose of the Major Rehabilitation Program "is to allow accomplishment of significant, costly, one-time structural rehabilitation or major replacement work (other repairs related to dam safety are accomplished under the normal O&M program)." [40] This program does not apply to facilities that were turned over to local interests for operation, maintenance, and major replacements after they were constructed by the Corps. The Dam Safety Assurance Program, however, applies to all dams built by the Corps regardless of current ownership. Specifically, this program "provides for modification of completed Corps dam projects which are potential safety hazards in light of current engineering standards and criteria.... This program is intended to facilitate upgrading of those project features which have design or construction deficiencies."[41] The Corps schedules rehabilitation under all of these programs based on funding availability.

U.S. Department of the Interior, Bureau of Reclamation

After Reclamation's engineers conduct dam safety inspections, through the Safety Evaluation of Existing Dams (SEED) account, any corrective action is carried out through the Initiate Safety of Dams Corrective Action program (ISCA).[42] Authority for Reclamation's dam safety program originates in the Reclamation Safety of Dams Act of 1978 (P.L. 95-578) and 1984 amendments (P.L. 98-404). Through ISCA appropriations, Reclamation focuses funding on priority structures "based on an evolving identification of risks and needs." Prior to recent legislation, Reclamation needed to submit a report to Congress for approval of modifications exceeding $750,000. The 108th Congress increased that ceiling to $1,250,000.[43] Costs incurred due to "age and normal deterioration of the structure" are considered normal operating costs and are cost-shared, with rates depending on the purposes for which the structure was constructed (project purposes). Furthermore, "modifications resulting from new hydrologic or seismic data or changes in the state of the art criteria" are cost-shared, with 15% of the costs allocated for reimbursement according to project purposes.

Rehabilitating Nonfederal Dams

In 2002, a task committee of the Association of State Dam Safety Officials estimated that $36.2 billion was needed to rehabilitate nonfederal dams and that $10.1 billion was needed by 2014 for repairs to "the nation's most critical dams."[44] Responsibility to undertake this rehabilitation generally falls to dam owners.

State Rehabilitation Funding

While the federal government does provide some direct funding for rehabilitating nonfederal dams, states and dam-owners bear the brunt of the costs. Most states have little funding to repair their own dams, let alone those dams owned by local governments, public utilities, and private entities. According to the task committee of the Association of State Dam Safety Officials, in 2003, nine states (AZ,MA, MD, NJ, NY, OH, PA, UT, and WI) have loan or grant programs to repair unsafe dams. These programs generally focus on publicly owned dams. Additional information on these programs is provided in the Task Committee's 2003 report and is recreated in **Table 3.**[45]

Table 3. State Grants for Dam Rehabilitation as of October 2003

State	Program Type	Program Name	Source and Amount of Funding	Eligibility	Loan/Grant Amount	Term of Loan
AZ	Loan or grant	Dam Repair	Legislature, lien fund, inspection fees, filing fees, principal and interest from previous loans	State engineer determnes dam to be dangerous to life, noneme-rgency	Loan for the cost of the project, or grant for a portion of project costs	Term of loan is up to 20 years at 3-6% interest depending on the length
MD	Loan and planning assistance	Maryland Environmental Service	State agency/ non-profit corporation	Counties, utilities and private groups. Need to have established service district for water supply, resource reclamation, dredging or storm water		
MA	Grants		Funding via Dept. of Environmental Management. No new appropriation as of Oct. 2003.	Local communities for repairs or removal	75% of the project, local share can be in-kind contributions	
NJ	Revolving loan fund New grant fund for municipally owned dams	Dam Restoration and Clean Water Trust Fund	\$20M-\$5M for state high hazard dams, \$15M loans. In 2000 an additional \$9.5[a] was added	Local units of governments, private owners can be co-applicants	Cost of project for loans Up to 100% for grants	Up to 20 years at 2% assessed against real estate benefitted
NY	Grants	Clean Water/Clean Air Bond Act	\$17M bonding	Municipality for dam safety projects	75% of eligible project (25% local match) \$300,000 cap per project	

Table 3. (Continued)

State	Program Type	Program Name	Source and Amount of Funding	Eligibility	Loan/Grant Amount	Term of Loan
OH	Revolving loan fund	Ohio Water Development Authority	Revolving loan fund	Owner must be under mandate from ODNR. *Dam Safety Loan Program*-local units of government, state districts. *Dam Safety Linked Deposit Program*-private ownership/orgs.		5-25 years at lower than market rate
PA	Revolving loan fund	Pennvest	Revolving loan fund, $2B from state general purpose funds	Projects associated with wastewater, water supply or stormwater	Up to cost of project	20-30 years at low interest
UT	Loans or grants	Utah Board of Water Resources	$4.5[a] from general revenue and 0.8 cent sales tax (created originally to deal with flood control problems)	High hazard dam owners. Mandated repairs	80-95% grant for irrigation or water supply dams, loans or grants for other owners	
WI	Grants	DNR Municipal Dam Grant Program	$11.5M of bonding over 10 years. Currently fully subscribed	Local units of government and Lake Districts	50-50 grants up to a $200,000 maximum	

Source: Association of State Dam Safety Officials. *The Cost of Rehabilitating Our Nation's Dams: A Methodology, Estimate and Proposed Funding Mechanism* Appendix B (Dec. 2002, revised Oct. 2003). See [http://www. damsafetycoalition.org/ files/pdf/asdsoreport.pdf].

a. Units not included in source table.

Federal Funding

While the federal government does help fund improvements to state dam programs, little federal funding is available for rehabilitating nonfederal dams. FEMA and NRCS are the primary sources of federal support for nonfederal dam rehabilitation.

FEMA

As described above, the Department of Homeland Security, through FEMA, runs the National Dam Safety Program which offers training and other assistance to state dam safety programs, though it generally does not provide rehabilitation funding. Legislation in the 109th Congress would have extended the NDSP and authorized funding through FY2011 and would have provided a total of $350 million to FEMA for a grant program for states to rehabilitate and repair publicly owned dams. Authorized funding would have been $50 million for FY2007 and $100 million annually through FY2011. Through other programs, FEMA may provide assistance to reduce the flood damage a failure would cause.[46] However, this assistance generally does not include funding for rebuilding or rehabilitating dams.

NRCS

Manyupstream flood control dams were built with assistance from the NRCS and turned over to local sponsors. Many of these dams are now nearing the end of their design life and some have significant rehabilitation needs. The NRCS has estimated that more than $540 million is needed to rehabilitate these dams.[47] In 2000, Congress enacted legislation authorizing $90 million in discretionary funding over five years to rehabilitate aging flood-water retention projects.[48] Funds are to be allocated based on an annual ranking of requests for rehabilitation assistance. For FY2002, Congress appropriated $10 million for this dam rehabilitation program.

For FY2003, the Administration did not include the program in its budget request. However, the 2002 farm bill[49] established the Small Watershed Rehabilitation Program, and included $275 million in mandatory funding for the program through FY2007. The Small Watershed Rehabilitation Program is significant because it is the first federal initiative to dedicate funding assistance to repair nonfederal dams (rather than tear down or build new dams). It uses a cost-share formula providing for 65% federal-35% local funding. In the past, cost sharing has been an equal responsibility of the federal government and local sponsors for most purposes.

Other Agencies

In general, federal agencies such as the Corps and Reclamation do not rehabilitate nonfederal dams. As described above, however, the Corps does sometimes repair structural deficiencies that resulted from its participation in the design or construction of the nonfederal dam. Though quite rare, Congress has also directed federal agencies to use their dam safety appropriations to rehabilitate a nonfederal dam that it did not help design or build.[50] However, representatives from the Corps and Reclamation have testified that they do not support such congressional direction.[51]

CONCLUSIONS

As the nation's dams age and development continues in floodplains, the structural integrity of this infrastructure will become a more significant public safety issue. In the aftermath of Hurricane Katrina, dams' planned capacity to withstand floods and other natural disasters has come under increased scrutiny. However, it is unclear to what extent there will be a widespread re-evaluation of flood and earthquake ratings at high-hazard dams. Such an evaluation could raise additional policy questions. For example: (1) what criteria should be used to determine whether current risks are acceptable; (2) if risks are not acceptable, should the dam be improved, or should other activities (e.g., changes to the design and or placement of downstream development) be undertaken; and (3) who should pay?

Regardless of whether dams were constructed to withstand an earthquake or flood of "appropriate" magnitude, they may have age-related deficiencies that need to be corrected to maintain current levels of safety. Therefore, it is likely that appropriations requests for safety inspections and rehabilitation activities will continue and may increase. It should be noted, however, that there currently are no clear criteria for prioritizing dam rehabilitation funding across agencies.

It is also unclear to what extent the federal government will fund inspection and rehabilitation activities at nonfederal dams. Through the National Dam Safety Program, the federal government provides training and assistance to state dam safety programs; authorization of appropriations for this program was extended through FY2011. Further, Congress has authorized appropriations for rehabilitation activities at several nonfederal dams. While there is likely to be an increasing demand for such assistance, there is currently no federal policy that describes the conditions under which federal funding is appropriate, nor has Congress established criteria for prioritizing funding among nonfederal projects.

End Notes

[1] This report was originally written by Kyna Powers with assistance from Richard Sachs.

[2] In 1972, failure of a mine tailings dam at Buffalo Creek, WV, flooded a 16-mile valley and killed 125 people. In 1976, Teton Dam (ID) failed, causing $1 billion in property damage and killing 11 people. Kelley Barnes Dam (GA) failed in 1977 killing 39 people and causing an estimated $2.8 million in damage. More recently, the 2003 failure of the Silver Lake Dam (MI) caused more than $100 million in damage, and the Kaloko Dam (HI) failed in March 2006, killing 7 people.

[3] The NID was first authorized in the National Dam Inspection Act of 1972 (P.L. 92-367) and later consolidated under the Water Resources Development Act of 1996, National Dam Safety Program Act (P.L. 104-303, Title II §215; 33 U.S.C. §467).

[4] The NID was last updated in February 2005. See [http://crunch tec.army.mil/nidpublic/webpages/nid.cfm].

[5] Other sources cite different figures; the online NID data is used throughout this report unless otherwise specified. When appropriate, these data are referred to as *NID Data.*

[6] One acre-foot of water is the amount of water that will cover an acre of land to a depth of one foot, approximately 326,000 gallons.

[7] *NID Data*.

[8] U.S. Dept. of Homeland Security, Federal Emergency Management Agency, *The National Dam Safety Program Fiscal Years 2000-2001* (Washington, DC: December 2001), p. 6 (Hereafter, cited as *FEMA report*, 2000-2001).

[9] P.L. 566 (83rd Cong.).

[10] U.S. Dept. of Agriculture, Natural Resources Conservation Service, *A Report to Congress on Aging Watershed Infrastructure* (Washington, DC: June, 2000), pp. 1-2. Hereafter cited as *NRCS Report*, 2000.

[11] *FEMA Report*, 2000-2001, p. 8-9. This figure differs from the NID-based data on dam ownership shown in **Table 2**. Some of these dams may not qualify for inclusion in the NID.

[12] Interagency Committee on Dam Safety. *Federal Guidelines on Dam Safety: Hazard Potential Classification System for Dams (FEMA 333).* (October 1998, reprinted January 2004).

[13] U.S. Dept. Of Homeland Security, Federal Emergency Management Agency. Dam Safety and Security in the United States: *A Progress Report on the National Dam Safety Program in FY 2002 and 2003 (FEMA 466)*, April 2004.

[14] National Performance of Dams, Dam Incidents Statistics Calculator, at [http://npdp.stanford.edu/index.html]. This database provides a low estimate of dam safety incidents, since reporting is voluntary; few private or local dams are included.

[15] Eugene P. Zeizel and Robert H. Dalton, "Agingof Dams and Urban Development Require Major Dam Rehabilitation Efforts: A Growing Problem in Floodplain Management," presented at the Annual Conference of the Association of State Flood Plain Managers, Charlotte, NC (June 3-8, 2001). This report indicated that 85% of all U.S. dams (not just NID dams) will be more than 50 years old by 2020.

[16] Association of State Dam Safety Officials, *State by State Statistics on Dams and State Safety Regulation , 2004.* See [http://www.damsafety.org/documents/Word/2004Stats.doc]. In 2003, FEMA reported that there were more than 2,600 unsafe dams based on a report by the American Society of Civil Engineers entitled *The 2003 Progress Report for America's Infrastructure* (September 4, 2003).

[17] Raul F. Silva, "A Methodology and Estimate of the National Cost for Dam Safety Rehabilitation," presented at a conference of the Association of State Dam Safety Officials, Providence, RI (September 27, 2000).
[18] Interagency Committee on Dam Safety and U. S. Dept. of Homeland Security, Federal Emergency Management Agency, *Federal Guidelines for Dam Safety* (June 1979 reprinted April 2004).
[19] While the Bureau of Reclamation manages the most dams, the U.S. Army Corps of Engineers manages the most NID dams.
[20] Alabama has 1,403 dams and Delaware has 61 dams.
[21] Association of State Dam Safety Officials, *State by State Statistics on Dams and State Safety Regulation — 2004.*
[22] Ibid.
[23] Average of reported state budgets; excludes Delaware, Louisiana, Michigan, Nevada, South Carolina, and South Dakota, which did not report their dam safety budgets. Association of State Dam Safety Officials, *State by State Statistics on Dams and State Safety Regulation — 2004.*
[24] Ibid.
[25] National Dam Safety Program Act, P.L. 104-303, Title II §215 (33 U.S.C. §467).
[26] P.L. 107-310, ICODS' authority is limited to coordination among federal entities only.
[27] National Dam Safety Program Act, P.L. 104-303, Title II §215 (33 U.S.C. §467)
[28] Association of State Dam Safety Officials, *Model State Dam Safety Program* (1998).
[29] Association of State Dam Safety Officials, *State By State Statistics on Dams and State Safety Regulation — 2004.*
[30] *FEMA Report*, 2000-2001, pp. 8-9. Of these 1,775 are included in the NID (261 significant-hazard and 697 high-hazard dams).
[31] FERC, Order 122, 46 *Fed. Reg.* 9036 (January 28, 1981), 18 C.F.R. §12.
[32] U.S. Dept. of Energy, Office of the Inspector General, *FERC Dam Safety Program,* DOE/IG 0486, (October 2000).
[33] P.L. 91-173, as amended by P.L. 95-164 (30 U.S.C. §801). See [http://www.msha.gov/REGS/ACT/ACTTC.HTM].
[34] Discussion with Kelvin Wu, Chief, Mine Waste & Geotechnical Engineering Division, Mine Safety and Health Administration, Washington, DC, on November 19, 2004.
[35] §313 of the Grain Standards and Warehouse Improvement Act of 2000 (P.L. 106-472).
[36] §2505 of the Farm Security and Rural Investment Act of 2002 (2002 Farm Bill).
[37] E-mail correspondence with Mr. Stuart Simpson, National Watersheds Program Leader, U.S. Dept. of Agriculture, NRCS, Washington, DC (Oct. 30, 2007).
[38] U.S. Dept. of Agriculture, *2009 Budget, Explanatory Notes for Committee on Appropriations*, vol. 2, pp. 18-36.
[39] Army Corps of Engineers, *Dam Safety Preparedness*, EP1110-2-13 (June 1996), p. 5-1.
[40] Ibid.
[41] Ibid.
[42] FY2009 Bureau of Reclamation Budget Justification, pp. BW-8 and 9.
[43] A Bill to Authorize Additional Appropriations for the Reclamation Safety of Dams Act of 1978, P.L. 108-439.
[44] A Task Committee of the Association of State Dam Safety Officials, *The Cost of Rehabilitating our Nation's Dams: A Methodology, Estimate, and Proposed Funding Mechanisms* (December 2002, revised October 2003), p. 4. See [http://www.damsafety coalition.org/files/pdf/asdsoreport.pdf].
[45] Ibid.
[46] See FEMA's Mitigation Grant Programs at [http://www.fema.gov/government/grant/index.shtm].
[47] *NRCS Report*, 2000, pp. 1, 15.
[48] The Grain Standards and Warehouse Improvements Act of 2000, P.L. 106-472, §313.

[49] §2505 of the Farm Security and Rural Investment Act, P.L. 107-171.

[50] See, for example, Goshen Dam (VA). Information from Charles Pearre, U.S. Army Corps of Engineers, Washington, DC, on February 2, 2005.

[51] U.S. Senate, Energy and Natural Resources Subcommittee on Water and Power, *Miscellaneous Water and Hydroelectric Project Bills,* S.Hrg. 108-271 (108th Congress), p. 28.

In: Water Infrastructure
Editor: Harald Nystroem

ISBN: 978-1-60692-479-2

Chapter 2

THE BUREAU OF RECLAMATION'S AGING INFRASTRUCTURE*

Nic Lane

SUMMARY

The Bureau of Reclamation (Reclamation) is responsible for the construction of most of the large irrigation and water resources infrastructure in the West. Reclamation manages water resource facilities in 17 western states with an original development cost of over $20.0 billion. Reclamation is over 100 years old, and many of its facilities now have an average age of over 50 years. This aging infrastructure requires increased maintenance and replacement efforts and expenditures.

Reclamation appears to have a well documented plan to assess the management needs of its portfolio of aging infrastructure. However, water resources infrastructure management objectives require prioritization due in part to a finite budget. This inevitably leads to conflicts over project funding priorities. As Reclamation's portfolio of infrastructure continues to age, these conflicts are likely to arise more often. Further, in the case of structures whose operations and maintenance responsibilities have been transferred to the local project beneficiaries, the increased maintenance and replacement efforts and

* This is an edited, reformatted and augmented version of a CRS Report for Congress publication dated April 2008.

expenditures associated with aging infrastructure — such as more expensive recapitalization projects — could motivate the beneficiaries to seek federal funds and congressional support to address project funding issues.

This report describes Reclamation's approach to managing aging infrastructure as well as that of two other agencies — the Army Corps of Engineers and the Natural Resources Conservation Service — involved with significant portfolios of dams and related infrastructure. Additionally, there is discussion of four specific approaches to managing Reclamation's aging infrastructure through legislative action.

Most large dams and water diversion structures in the West were built by, or with the assistance of, the Bureau of Reclamation (Reclamation). Reclamation's mission is to manage, develop, and protect water and related resources in an environmentally and economically sound manner in the interest of the American public.[1] In most cases this means developing water supplies primarily for irrigation to reclaim arid lands in the West. Reclamation manages water resource facilities in 17 western states with an original development cost of over $20.0 billion.[2] Citing engineering indices, the agency has estimated that the replacement value of all Reclamation infrastructure would be over $100 billion.[3] Reclamation's inventory of assets includes 471 dams and dikes that create 348 reservoirs with a total storage capacity of 245 million acre-feet of water.[4] Reclamation is the largest wholesaler of water in the United States. Through its assets, Reclamation serves more than 31 million people, and provides irrigation water for 10 million acres of farmland that produce 60% of the nation's vegetables and 25% of its fruits and nuts. Reclamation is the nation's second-largest producer of hydroelectric power, with 58 hydroelectric power plants that provide an average of more than 42 billion kilowatt-hours of energy each year.

Reclamation has over 77% of the Department of the Interior's (DOI's) constructed assets, making the agency an important manager of infrastructure and supporting assets. Reclamation is over 100 years old, and many of its facilities now have an average age of over 50 years. This aging infrastructure requires increased maintenance and replacement efforts and expenditures.[5] Reclamation's limited budget for asset management could lead to conflicts over project funding priorities. (See discussion of St. Mary Dam, below.) As Reclamation's infrastructure continues to age, these conflicts may arise more frequently and could become an issue for Congress. S. 2842, the Aging Water Infrastructure Act, addresses this issue and would require that the Secretary of the Interior, acting through the Commissioner of Reclamation, carry out annual inspections of water resources infrastructure for Reclamation-owned and -operated facilities, as well as for those facilities owned by Reclamation, but operated and maintained by water users. Additionally, the bill would require the Secretary to develop a national priorities list of infrastructure maintenance needs and establish standards and guidelines for the maintenance of these facilities.

Reclamation estimates that approximately $3 billion will be required to rehabilitate, modify, or replace water resources assets in the future. This estimate was cited as only a rough approximation of conceivable need based on data that are not reliable for budgeting or long-term planning decisions. Reclamation estimates that at least $800 million of the $3 billion need is for transferred works.[6] Transferred works arethosefacilities owned by Reclamation but operated and maintained by end users such as irrigation districts. That would leave approximately $2.2 billion needed for reserved works, which are facilities both owned and operated by Reclamation. Approximately two-thirds of Reclamation facilities are transferred facilities.[7]

Reclamation states it is preparing a document summarizing the condition of its aging infrastructure, historical and current procedures to address these identified needs, and available data on the condition of its infrastructure. The document is not available for distribution outside the executive branch yet; it is still undergoing internal review.[8]

As a comparison, the Army Corps of Engineers (Corps), which operates and maintains a substantial inventory of dams and other water control infrastructure throughout the country, indicates that in FY2006 it had $1.8 billion in deferred maintenance for its civil works activities.[9]

Reclamation's Approach to Aging Infrastructure Management

General Approach

Maintenance, repair, and rehabilitation needs at Reclamation projects are typically identified through regular monitoring by project operators, or more formal facility condition assessments performed by technical experts. Maintenance activities are scheduled, funded, and accomplished based on an assessment of the work's necessity.[10] Abrupt facility failures may receive immediate maintenance action to restore facility service or protect public safety. Finally, if the facility has reached or exceeded its expected service life, Reclamation may consider a major rehabilitation or replacement activity (i.e., recapitalization effort).[11]

The primary day-to-day responsibility for asset management lies with Reclamation's 25 area offices. Each office reports to one of the five Reclamation regional offices which, in turn, reports to the Deputy Commissioner for Operations. Ultimately, asset portfolio decisions are the responsibilityof the Commissioner, who relies on several advisory bodies for assistance.[12]

Reclamation's Facilities Operation and Maintenance (O&M) Team addresses Reclamation-wide O&M-related priorities, issues, activities, program and budget formulation, and facilitate program accomplishment. The team's responsibilities include reviewing and making recommendations to the Deputy Commissioner for Operations on issues including:

- deferred maintenance,
- asset management,
- condition assessments/field review activities,
- Replacements, Additions, and Extraordinary Maintenance (RAX) items,
- facility security and public safety, and
- maintenance management practices and systems.[13]

The identification of maintenance needs is similar for both transferred works and reservedworks. Maintenanceneeds are identified either by the operatingentities or through periodic facility reviews conducted by Reclamation.

Major maintenance or rehabilitation for reserved works is typically funded from annual appropriations.[14] However, maintenance funding for transferred works is the responsibility of the operating entity, which may have difficulty funding major recapitalization efforts. Securing needed additional funding through private lenders for rehabilitation work on transferred facilities can be difficult because the federal government retains title, so the facilities cannot be used as collateral by the operators to secure a loan. In FY2008 Reclamation requested funding for a loan guarantee program to aid operating entities in their pursuit of private funds.

Specific Cases

In some instances the rehabilitation and maintenance of specific Reclamation dams and water projects are addressed directly by legislation. These approaches address individual facilities (the Reclamation Safety of Dams Program being an exception), but may have broader effects on public policy — for example, if proponents of future rehabilitation projects view them as creating precedents or appearing to provide a number of different policy models to choose from. Four cases are discussed below.

Arrowrock Dam

Arrowrock Dam was completed in 1915. The 350-foot high concrete arch dam is about 13 miles east of Boise, Idaho. Water releases at the dam are controlled by 20 ensign valves and five sluice gates. After nearly 90 years of continuous operation, Reclamation replaced all 10 of the lower valves with devices called clamshell gates that would permit inspection and maintenance without the need to draw down the level of the reservoir.[15]

Reclamation worked over a three-year period to install the clamshell gates that replaced the 89-year-old ensign valves. Total cost for the project, which was completed in the summer of 2004, was about $20 million of federal funding. The project required coordination between Reclamation and irrigation districts, fishery biologists, recreation interests, and the Corps to manage flows during construction in order to ensure adequate water flows for all interests.[16]

In §206 of the Energy and Water Development Appropriations Act of 2002,[17] Congress stipulated that Reclamation was to recover no more than $6.9 million (35%) of reimbursable expenses typically charged to water users for project O&M costs at Arrowrock Dam. Congress further stipulated that these costs were to be recovered over a 15-year period. Reclamation typically requires O&M costs to be repaid by project beneficiaries within the fiscal year that the funds were disbursed.[18]

Jackson Gulch Dam

The Mancos Project in southwestern Colorado includes the Jackson Gulch Dam and reservoir, as well as inlet and outlet canals. It provides supplemental irrigation water for nearly 14,000 acres. The project was approved by President Roosevelt on October 21, 1940.[19] Construction began in 1941, and was completed nine years later. The first water from Jackson Gulch Reservoir was delivered to water users in 1949. Responsibility for O&M at the project was transferred to the Mancos Water Conservancy District in 1963.[20]

S. 1477, the Jackson Gulch Rehabilitation Act, would direct the Secretary of the Interior, through Reclamation, to pay an 80% share of the cost of the Jackson Gulch Rehabilitation Project. The bill would authorize $6.45 million for the federal share of rehabilitation work. The Mancos Water Conservancy District would remain responsible for O&M at the project. Reclamation's FY2009 budget request for facility maintenance and rehabilitation at the Mancos project is $64,000.[21] This amount is less than 1% of the funding authorized for rehabilitation, thus it appears that the agency is not yet seeking funding to begin rehabilitation work at the project.

Reclamation does not support S. 1477, stating that the activities authorized by the bill are the responsibility of the Mancos Water Conservancy District. Reclamation considers rehabilitation work to be O&M, and thus the responsibility of the water users.[22]

S. 1477 would provide a legislative response to a specific conflict. That conflict centers around the federal role in maintaining the nation's aging infrastructure versus adherence to interpreted contractual obligations and reliance on the ability of project stakeholders to perform the more significant work associated with rehabilitation.

Dam Safety Program

The Reclamation Safety of Dams Act of 1978[23] gave the Secretary of the Interior permanent authority to modify Reclamation dams and related facilities to ensure their structural integrity and safety. The act provides direction on how repair or upgrade work is to be classified for repayment. Generally, repair costs that are the result of modifications due to age, normal deterioration, or nonperformance of normal maintenance are considered reimbursable project costs, while those resulting from new safety criteria or new hydrologic or seismic data are nonreimbursable. Reclamation's Dam Safety Program has two primary components, the Safety of Dams Evaluation and Modification Program and the DOI Dam Safety Program. The DOI Dam SafetyProgram is a relatively small program — the FY2009 budget request is $1.25 million — that provides facilitation and guidance to other DOI departments on their dam safety programs.

The Safety of Dams Evaluation and Modification Program constitutes the bulk of the dam safety budget, with an FY2009 budget request of $90.0 million. It focuses specifically on Reclamation dams and has two primary sub-tasks, the Safety Evaluation of Existing Dams program (SEED) and the Initiate Safety of Dams Corrective Action program (ISCA). The SEED program is focused on the analysis and identification of potential hazards or increased risk at Reclamation dams, while ISCA is the implementation component of the program aimed at the study, identification, and accomplishment of repairs or rehabilitation.

When a need for corrective action is identified, that action is funded under ISCA. Funding requests to begin and continue the construction of remedial actions identified through ISCA are transferred from the Dam Safety Program and assigned to the specific project.

Work funded through ISCA is conducted under authority granted by the Safety of Dams Act; however, the SEED program is a response to the October 4, 1979, presidential memorandum directing federal agencies to implement the *Federal Guidelines for Dam Safety*.[24] SEED work is considered a public benefit and its costs are not assigned to a specific project for reimbursement.[25]

St. Mary Dam and Diversion

The St. Mary Diversion Dam is part of Reclamation's MilkRiver Projectinnorthwestern Montana. The dam, located on the St. Mary River, is 6 feet high and diverts water to the St. Mary Canal. The canal runs 29 miles to the point where the water is discharged into the North Fork of the Milk River.[26] Features of the canal in need of repair are sections of the earthen canal itself; two 90-inch wide, 3,600-foot long steel pipes; a second set of pipes, 78 inches in diameter and 1,405 feet long; and a series of five large concrete drops at the lower end of the canal. The facility is owned and operated by Reclamation.

The canal is designed to carry 850 cubic feet of water per second (cfs); however, the current capacity is reported to be 620 cfs, nearly 30% lower. It is reported that the earthen canals are crumbling, the pipes that carry the water across the St. Mary River and Hall's Coulee are affected by slope instability and leakage, and the concrete drops at the end of the canal are cracking.[27]

The Water Resources Development Act of 2007 (WRDA),[28] §5103, authorizes $153 million for the Secretary of the Army, in consultation with Reclamation, to study, plan, design, and construct the rehabilitation of the St. Mary Dam and Conveyance Works. Except for construction associated with standard O&M and for emergency repairs that ensure water transportation or protect life or property, no construction is authorized to begin until January 2011, or earlier if Congress approves the water rights settlement of the Blackfeet Tribe and Fort Belknap Indian Community. Reclamation has requested $135,000 for facility maintenance and rehabilitation for all of the infrastructure that comprises the Milk River Project in FY2009.[29]

The 109th Congress proposed legislation (S. 3563) that would have authorized Reclamation to conduct feasibility studies on the rehabilitation of St. Mary facilities. The bill would also have authorized planning, design, and construction for rehabilitation work on the St. Mary facilities. The repayment allocation indicated was 55% nonreimbursable and 45% reimbursable. Additionally, the federal share of the reimbursable construction costs was to be 75%. Thus, the total federal share would have been 88.75%.

Reclamation indicated support for the proposed feasibility studies in S. 3563 but did not support authorization for construction in WRDA or the cost-sharing arrangement in the bill. The agency indicated that authorization of construction prior to the completion of feasibility studies was premature, and that pending tribal water rights claims should be resolved prior to rehabilitation work at the project. Further, Reclamation opposed the proposed project repayment terms, citing a concern that repayment terms which depart from standard Reclamation practice are precedent-setting and may pose problems for the agency in the future.[30] Typically, Reclamation has constructed projects using federal funds and then established a repayment schedule for project beneficiaries based on the amount of total construction costs allocated to various project purposes. Reclamation project authorizations typically require 100% repayment, with interest.[31]

CORPS AND NRCS APPROACHES TO AGING INFRASTRUCTURE MANAGEMENT

Two other federal agencies associated with significant inventories of dams or other water resources infrastructure are the Corps of Engineers (Department of Defense) and the U.S. Department of Agriculture's Natural Resources Conservation Service (NRCS). They have different roles and responsibilities regarding O&M and rehabilitation. The Corps is an active federal dam operator, while NRCS was a partner in the construction of many agricultural dams that are the responsibility of local entities, and is now a partner in their rehabilitation. The O&M prioritization programs for the Corps and NRCS are summarized below.

The Corps of Engineers[32]

The Corps is responsible for water resources projects that provide flood damage reduction, navigation, hydropower generation, and water supplies.[33] The agency operates and maintains 609 dams, including 75 hydroelectric power stations that generate 24% of the nation's hydropower and 3% of its total electricity. It also operates and maintains 257 navigation locks.[34]

For each budget cycle, the Corps publishes its procedures for ranking O&M actions for the portfolio of its infrastructure. The budget process starts at the Corps district and division level,[35] where individual budget activities within business lines are identified. Business lines are functional categories such as hydropower, navigation, and water supply. Each activity is ranked based on the O&M procedures for its business line for that budget cycle, and the cost of each activity is estimated. The rankings are largely determined by project performance measures; most of these measures are based on economic (e.g., tons of commercial cargo moved, benefit-cost ratio), safety, or environmental outputs. The smallest increment of work for O&M funding is a work item or task (e.g., mowing a levee or painting a lock gate). Since the adoption of a performance-based budgeting approach, the Administration's most recent budget requests for O&M fundinghavebeen organized by geographic regions or basins, rather than for individual line items. As indicated by the Corps' budget development procedures, the process for developing the basin or region O&M estimates consists of aggregating estimates for individual actions.

The actions included in the Administration's final budget request are determined by Corps headquarters in conjunction with the Office of Management and Budget based on the rankings supplied from the Corps Divisions, as long as they are consistent with established Corps and Administration policy (which may include the Administration's overall level of support for the O&M account) and adhere to the procedures established for that budget cycle.[36] Although the Administration identifies and prioritizes project O&M funding needs based on Corps procedures and Administration policy, Congress also directs Corps funding during its annual consideration of Energy and Water Development appropriations bills.

Natural Resources Conservation Service

The NRCS has participated in the construction of over 11,000 dams. Local sponsors own the dams and are responsible for their operation and maintenance. In 2000, the Watershed Protection and Flood Prevention Act[37] was amended to authorize NRCS to assist with rehabilitation of these aging dams. The NRCS approach to aging infrastructure is driven by the end user, with assistance from the NRCS Watershed Rehabilitation Program allocated based on the risk to public safety and availability of appropriations.

NRCS staff indicate that they do not make estimates of future rehabilitation funding needs.[38] Rehabilitation projects are identified by local entities as they apply for rehabilitation funding assistance. Local sponsors can apply for the Watershed Rehabilitation Program at any time as long as they are within the eligibility criteria for the program. When NRCS receives an application, NRCS's state conservationists perform a risk assessment of the hazards associated with the dam. The state conservationists forward assistance requests along with data from the risk assessment to NRCS headquarters annually. The national office evaluates all requests and funds those that pose the highest risk to public safety within available appropriations. The amount funded by NRCS is typically less than the total applications for aid,[39] and total applications for aid do not necessarily reflect the total need. For FY2003 through FY2008, authorized funding for the program was at the following levels: FY2003, \$29.8 million; FY2004, \$29.6 million; FY2005, \$27.5 million; FY2006, \$31.5 million; FY2007, \$31.3 million; and FY2008, \$19.9 million. The Administration's FY2009 budget request for this program is \$5.9 million. All authorized funding is subject to appropriations via the Agriculture appropriations bill.[40]

SUMMARY AND ANALYSIS

There is an established process within Reclamation for identifying and prioritizing the O&M or rehabilitation funding needs of the water resources infrastructure that Reclamation has constructed since its creation in 1902. However, it is clear that these needs are addressed in a number of ways both internal and external to Reclamation — i.e., through the Dam Safety Program and through congressionally directed spending.

As the dams and water conveyance structures continue to age and need more costly repairs and upgrades, the question of who should pay will likely be raised repeatedly. For Jackson Gulch (the Mancos Project), even if the contractual responsibility to conduct rehabilitation work is clearly defined as O&M and remains the responsibility of water users, are the nation's best interests served by allowing infrastructure to deteriorate further when those water users are unable or unwilling to undertake repairs? Or should water users be held accountable for their contractual obligations? If the latter, should funding for, and transfer of, future projects be contingent on stricter analyses of end users' ability to repay their obligations?

In contrast to S. 1477 and the Jackson Gulch Project described above, at Arrowrock Dam the approach to the rehabilitation of aging infrastructure stipulated by Congress in §206 of the Energy and Water Development Appropriations Act of 2002 appears to classify the replacement of worn-out equipment as O&M. The repayment of these costs is typically the responsibility of water users, to be repaid in the fiscal year that costs are incurred. However, the legislation caps the total responsibility of water users at $6.9 million and extends the repayment period to 15 years.

For the St. Mary Dam issue, Congress chose the approach of designating— and authorizing funding for — another federal agency to do the work instead of Reclamation. However, the solution may add complexity to an already complicated issue. The WRDA legislation stipulates a 75% federal cost share of the total cost of the rehabilitation project, but does not specify repayment of the remaining 25%. It might be assumed that the remaining cost is to be borne by water users under Reclamation's repayment rules, but this is not specified. Also, this cost share allocation may conflict with §4(a) of the Reclamation Safety of Dams Act of 1978,[41] which indicates that costs associated with modification of structures resulting from age and deterioration or nonperformance of normal maintenance by the operating entity are to be considered reimbursable project costs.

Further, regarding St. Mary Dam, the WRDA stipulation for an exclusion on construction associated with standard O&M as well as that for emergency repairs[42] appears to indicate that the broader St. Mary rehabilitation project is considered neither routine O&M nor an emergency repair. If that is the case, it may bolster the position of those who would seek to classify the work as "modification of structures resulting from age and deterioration ornonperformance of normal maintenance" cited from the Safety of Dams Act above, and thus subject to reimbursement under the terms of the act.

Finally, as its name suggests, the Dam Safety Program is focused solely on dams. Review, analysis, and repair of other water resources infrastructure such as canals and pump stations are not within the scope of this program. The

Dam Safety Program is clear in its design, with well-delineated evaluation and implementation components. However, its limited scope — albeit focused on a vital element of water infrastructure — excludes other important water resource facilities.

Based on a review of documents such as the *Bureau of Reclamation Asset Management Plan, Fiscal Year 2007*, commentors would likely conclude that the agency has a well-documented plan to assess the management needs of its portfolio of aging infrastructure. However, as may be indicated by some of the projects discussed above such as the St. Mary Dam, the necessity of prioritization given a finite budget for asset management objectives inevitably leads to conflicts over project funding priorities. As Reclamation's infrastructure continues to age, these conflicts may arise more frequently. Additionally, more expensive recapitalization projects may exceed the financial means of local operators in the case of transferred works and could drive those entities to seek congressional support for project funding.

End Notes

[1] See [http://www.usbr.gov/main/about/mission.html].

[2] U.S. Dept. of the Interior, Bureau of Reclamation, *Bureau of Reclamation Asset Management Plan Fiscal Year 2007,* pp. 5 and 10. Hereinafter referred to as "AMP."

[3] Ibid.

[4] One acre-foot is the amount of water required to cover an area of one acre to a depth of onefoot — 325,851 gallons.

[5] AMP, pp. 5 and 10.

[6] Testimony of Mr. Robert Johnson, Commissioner, U.S. Dept. of the Interior, Bureau of Reclamation, before the Senate Committee on Energy and Natural resources, Subcommittee on Water and Power, April 17, 2008.

[7] AMP, p. 7.

[8] Email correspondence on February 28, 2007, with Mr. Carter Brown, Special Assistant, Congressional and Legislative Affairs Office, U.S. Dept. of the Interior, Bureau of Reclamation, Washington, DC. Hereinafter referred to as "Brown, February 28."

[9] Dept. of Defense, *Performance and Accountability Report for Fiscal Year 2006*, p. 164. Available at [http://www.defenselink.mil/comptroller/par/fy2006/Entire_Document_(7.8_MB).pdf]. The 2007 edition of this report does not update this estimate, and the Corps has not provided CRS with a more recent figure.

[10] The guidelines Reclamation follows for infrastructure evaluations are indicated in the AMP, p. 37.

[11] Brown, February 28.

[12] AMP, p. 30.

[13] Ibid.

[14] Ibid.

[15] See [http://web.archive.org/web/20030813035609/www.usbr.gov/pn/programs/arrow rockvalve/].

[16] See [http://www.usbr.gov/newsroom/newsrelease/detail.cfm?RecordID=1582].

[17] P.L. 107-66.

[18] Document provided on April 18, 2008, by Reclamation's Contracts Services Office, Denver, CO; and testimony of Mr. Charles McGinnis, National Research Council Committee, before the Senate Committee on Energy and Natural Resources, Subcommittee on Water and Power, April 17, 2008.

[19] Water Conservation and Utilization Program Act of August 11, 1939 (53 Stat. 1418), as amended October 14, 1940 (54 stat. 1119).

[20] Reclamation website for the Mancos Project at [http://www.usbr.gov/dataweb/html/mancos.html].

[21] Bureau of Reclamation FY2009 Budget Justification, *Upper Colorado Region*, p. 49.

[22] Hearing before the Subcommittee on Water and Power of the Senate Committee on Energy and Natural Resources, testimony of Mr. Larry Todd, Deputy Commissioner for Policy, Administration and Budget, Bureau of Reclamation, S. Hrg. 110-152, *Miscellaneous Water and Power Legislation*, July 26, 2007, p. 9.

[23] P.L. 95-578, and amended in 1984 under P.L. 98-404.

[24] U.S. Dept. Of Homeland Security, Federal Emergency Management Agency, *Federal Guidelines for Dam Safety*, June 1979, reprinted April 2004. Available at [http://www.fema.gov/library/viewRecord.do?id=1578].

[25] Bureau of Reclamation FY2009 Budget Justification, *Bureauwide* pp. 8, 9, and 11.

[26] See [http://www.usbr.gov/dataweb/html/milkriver.html#general].

[27] Hearing before the Senate Committee on Energy and Natural Resources, statement of Senator Max Baucus, S. Hrg. 109-764, *St. Mary Diversion and Conveyance Works and Milk River Project*, September 1, 2006, p. 3; and Paul Azevedo, Montana Dept. of Natural Resources and Conservation, *The Need to Rehabilitate the St. Mary Facilities*, p. 1. See [http://dnrc.mt.gov/st_mary/pdfs/stmarybackground.pdf].

[28] P.L. 110-114.

[29] Bureau of Reclamation FY2009 Budget Justification, *Great Plains Region*, p. 51.

[30] Hearing before the Senate Committee on Energyand Natural Resources, statement of Mr. Mike Ryan, Regional Director, Bureau of Reclamation, S. Hrg. 109-764, *St. Mary Diversion and Conveyance Works and Milk River Project*, September 1, 2006, pp. 40-42.

[31] Two seminal laws shaping Reclamation policy on project repayment are the Reclamation Act of 1902, 43 U.S.C. § 391, and the Reclamation Project Act of 1939, 43 U.S.C. § 485.

[32] Nicole Carter, CRS Specialist in Natural Resources Policy, contributed to this section.

[33] See [http://www.usace.army.mil/missions/#Water%20Resources].

[34] See [http://www.usace.army.mil/missions/#Infrastructure].

[35] Within the Corps there are eight Divisions, or Major Subordinate Commands (MSC) across the United States. In turn, these are comprised of 41 Districts. Division and District boundaries are defined by watershed, not states.

[36] Dept. of Defense, U.S. Army Corps of Engineers, *Army Programs — Corps of Engineers Civil Works Direct Program — Program Development Guidance Fiscal Year 2009*, pp. C-2-13 and C-2-14.

[37] 68 Stat. 666; 16 U.S.C. §§ 1001-1008.

[38] Email correspondence with Mr. Stuart Simpson, National Watersheds Program Leader, U.S. Dept. of Agriculture, NRCS, Washington, DC (October 30, 2007, and February 21, 2008). Hereinafter referred to as *Simpson*.

[39] In FY2005, NRCS received approximately $46 million in aid applications and dispensed approximately $27 million in aid funding. U.S. Dept. of Agriculture, NRCS, *Watershed Rehabilitation: A Progress Report 2005*, p. 2 (January 2005).

[40] FY2003 through FY2007 data, *Simpson*. FY2008 and FY2009 data, U.S. Dept. of Agriculture, *2009 Budget, Explanatory Notes for Committee on Appropriations, vol. 2*, pp. 18-36.

[41] P.L. 95-578.

[42] P.L. 110-114, § 5103(c)(2)(A) and (B).

In: Water Infrastructure
Editor: Harald Nystroem
ISBN: 978-1-60692-479-2

Chapter 3

SENATE COMMITTEE ON ENERGY AND NATURAL RESOURCES. WATER AND POWER SUBCOMMITTEE HEARING: TESTIMONY ON THE INCREASING NUMBER OF ISSUES ASSOCIATED WITH AGING WATER RESOURCE INFRASTRUCTURE THAT ARE OPERATED AND MAINTAINED, OR OWNED, BY THE UNITED STATES BUREAU OF RECLAMATION (SD-366)

Wesley Randal "Randy" Reed

Mr. Chairman, distinguished members of the Committee, my name is Randy Reed. I am a fourth generation agricultural producer from northern Montana and serve as Co-Chair of the St. Mary Rehabilitation Working Group with Montana Lt. Governor John Bohlinger. I would like to thank the Committee for the opportunity to provide testimony concerning the critical issue of the Bureau of Reclamation's aging water resource infrastructure. The example I bring before the Committee is the Milk River Project and the associated St. Mary Diversion and Conveyance Works.

In 1903, Congress authorized construction of the Milk River Project as one of the first five reclamation projects built under the Reclamation Act of 1902. Development of the Milk River Project required construction of the St. Mary Diversion & Conveyance Works to divert and transport water from the St. Mary River Basin to the Milk River Basin.

The crisis we are facing today is the potential loss of the St. Mary Diversion & Conveyance Works. After over 90 years of service, many components of the system have exceeded their 50-year design life, and are in critical need of repair or replacement. Engineering investigations, an increase in the frequency and cost of extra-ordinary maintenance needs, and recurrent delivery interruptions indicate the system is on the verge of collapse.

In an average year approximately 70% to 90% of the water diverted from the Milk River for municipal and agricultural use originates in the St. Mary River Basin. Without this imported water, the Milk River would run dry six out of ten years. The stable supply of water provided by the system secures the "backbone" of the region's agricultural economy

For the last 7 years I have worked with the State of Montana and stakeholders in the Milk River Basin to engage the Bureau of Reclamation to develop a solution for rehabilitating the St. Mary Diversion and Conveyance Works before the system suffers catastrophic failure. However, rather than taking a leadership role, Reclamation officials have deferred to pressure by the Office of Management and Budget and initially resisted our efforts. Even today, they are playing a passive role in the rehabilitation of one of the nation's original reclamation projects.

Reclamation's position is confusing to contract holders and basin residents who do not understand why the federal agency that owns and operates the facilities is not leading the rehabilitation effort. Not only is it confusing for the owner/operator not to be fully engaged in the process, it has been detrimental to the overall effort.

Reclamation appears to be paralyzed in the face of the tidal wave of aging infrastructure issues they face across the West. It appears that without direction from Congress, the Department of Interior will not allow Reclamation to actively participate with project irrigators and the State of Montana to rehabilitate the St. Mary system. Through inaction at the federal level, the federal government is missing an opportunity to work with stakeholders and the State of Montana to find a workable solution for ensuring the continued viability of the Milk River Project. This federal inertia may also impact the federal government's ability to settle federal reserved water rights claims with the Blackfeet Tribe and Ft. Belknap Indian Community and jeopardize the ability of the United States to access water from the St. Mary River under terms of the 1909 Boundary Waters Treaty. Certainly, lack of leadership by federal agencies is a threat to my business and has forced me to get out and lead.

I believe that the Bureau of Reclamation must be given the ability to develop new tools for addressing the challenges presented by aging water resources infrastructure. One challenge we face in the Milk River Basin is how the Bureau of Reclamation views capital expenditures for rehabilitation projects. According to a 2006 engineering study undertaken by the state of Montana the cost to rehabilitate and replace the St. Mary Diversion & Conveyance Works is estimated at $130 million to $140 million. Reclamation considers such work to be part of operation and maintenance and requires the cost to be paid back in the year they are incurred. This presents an insurmountable financial obstacle to irrigators in the Milk River Project. Irrigators within our Project would benefit from the development of a program that enables water users to modernize or rehabilitate their projects and payoff those costs over time under reasonable terms and conditions. Another option would be to encourage partnerships of federal agencies with complementary resources and interests.

As originally authorized, the St. Mary Diversion & Conveyance Works are operated for the single purpose of irrigation. As such, nearly 100% of the cost to rehabilitate and replace the system must be borne by small irrigators, like me, within the irrigation districts holding water delivery contracts. Today the beneficiaries extend far beyond this original intent. The Bureau of Reclamation's 2005 "Current Use Benefits Analysis" showed large public benefits accrue from the existence of the St. Mary system. In addition to providing water for irrigated agriculture, the system provides water to communities and industry, water to the Bowdoin National Wildlife Refuge, and water that supports a wealth of recreational opportunities in north-central Montana. The system is also integral to settlement of federal reserved water rights with the Blackfeet Tribe and Ft. Belknap Indian Community, and implementation of the 1909 Boundary Waters Treaty. Amending the original authorization for the Milk River Project and St. Mary Diversion and Conveyance Works to reflect the extensive public benefits they provide would relieve irrigators of the burden of subsidizing the federal government for the benefits enjoyed by others.

My great grandfather homesteaded in the Milk River Valley and was among the founders supporting construction of the Milk River Project and St. Mary Diversion & Conveyance Works at the turn of the 20th century. Irrigation allowed my great grandfather to settle in Northern Montana and endure. Today, my family benefits from these same water resource facilities and we are able to raise irrigated certified seed potatoes, alfalfa hay and malt barley. We also rely on the project for our drinking water and enjoy the many recreational opportunities supported by the water system.

Failure of the St. Mary Diversion & Conveyance Works would economically devastate communities and businesses along the Hi-Line and likely have economic repercussions across the state. Aging infrastructure threatens my family's future, the future of the Milk River Basin, and the future of farm-based economies across the West.

I urge the Committee to remember that many of these Reclamation projects are federal assets owned and operated by the federal government. Investment in these projects will preserve our Nation's ability to conserve, enhance, and efficiently manage our most precious natural resource, WATER!

I would like to thank the Chairman, and the Committee for their recognition of the aging infrastructure problem and timely attention to finding a solution. Again, I appreciate this opportunity to testify.

INTRODUCTION

A century ago local and national leaders developed a vision to build an economy along the Hi-Line of north central Montana. Implementation of this vision required construction of the St. Mary Diversion & Conveyance Works to divert and transport water from the St. Mary River Basin to the Milk River Basin. Water imported from the St. Mary River provides supplemental water to the Bureau of Reclamation's Milk River Project.

Today this marvel of ingenuity and early 20th Century engineering is the keystone to the economic vitality of Montana's Hi-Line region. In addition to fulfilling the original intent of providing water for large-scale irrigated agriculture along the Milk River, the system also provides water to communities and industry, water to Bowdoin National Wildlife Refuge, and water that supports recreational opportunities for fishing and hunting along 650 miles of river and at two major reservoirs in north-central Montana. In short, the St. Mary Diversion & Conveyance Works is truly "the lifeline of the Hi-Line". The system is also integral to settlement of federal reserved water rights with the Blackfeet Tribe and Ft. Belknap Indian Community, and implementation of the 1909 Boundary Waters Treaty between the United States and Canada.

However, after 90 years of service, many components of the system have exceeded their 50-year design life, and are in critical need of repair or replacement. Engineering investigations, an increase in the frequency and cost of extra-ordinary maintenance needs, and recurrent delivery interruptions indicate the system is on the verge of collapse.

In an average year approximately 70% of the water diverted from the Milk River for municipal and agricultural use originates in the St. Mary River Basin. Without this imported water, the Milk River would run dry six out of ten years. The stable supply of water provided by the system secures the "backbone" of the region's agricultural economy. Failure of the St. Mary Diversion & Conveyance Works would economically devastate communities and businesses along the Hi-Line, and likely have economic repercussions across the state. Failure of the canal, siphons, or drop structures may also result in environmental damage on the Blackfeet Reservation and in southern Alberta. Loss of the system would also curtail the United States ability to access its share of water from the St. Mary River in accordance with the Boundary Waters Treaty of 1909.

BACKGROUND

On March 14, 1903, Secretary of Interior Ethan Allen Hitchcock authorized construction of the Milk River Project (Project) as one of the first five reclamation projects built by the newly created Reclamation Service (now Bureau of Reclamation) under the Reclamation Act of 1902. The Project's objective was to provide a stable source of water for irrigation of the lower Milk River valley, and settlers moved to the valley on that promise. Early settlers had learned that natural flows in the Milk River did not provide a reliable water source for irrigation in the downstream end of the watershed. Consequently, a plan to divert water from the St. Mary River to augment flows in the Milk River was a key component of the Milk River Project. In 1905, Secretary Hitchcock authorized construction of the St. Mary Diversion & Conveyance Works to provide supplemental water to the Milk River Project. Construction of the diversion dam and canal began on July 27, 1906. Diversion and delivery of supplemental water to the Milk River Project started in 1916.

The St. Mary Diversion & Conveyance Works are located entirely on the Blackfeet Reservation in Glacier County, Montana. The system annually diverts approximately 160,000 acre-feet of water from the St. Mary River east of Glacier National Park to the North Fork of the Milk River via a 90-year old, 29-mile long facility. Separate components include:

- Sherburne Reservoir – located on Swiftcurrent Creek just outside the eastern boundary of Glacier National Park;
- Swiftcurrent Creek Dike;
- Diversion dam and canal headgates on the St. Mary River;
- St. Mary River Siphon – Two 7.5 ft (dia.), 3,200 foot long riveted steel barrels with a combined discharge capacity of 850 cubic feet per second (cfs);
- Hall Coulee Siphon - Two 6.5 ft (dia.), 1,405 foot long riveted steel barrels;
- Check and wasteway structures – most are currently inoperable;
- Five hydraulic drop structures – combined vertical drop of 214 ft; and
- Approximately 29 miles of single bank, unlined, earthen canal.

The system is owned by the U.S. Government and is operated and maintained by the U.S. Bureau of Reclamation (Reclamation) with funds paid by holders of water delivery contracts in the Milk River Basin.

The St. Mary Diversion & Conveyance Works are approaching 100 years old, and are still dependent on the same basic infrastructure built by the Reclamation Service in the early 1900's. Continued degradation of the system has resulted in a loss of capacity from 850 cubic feet per second (cfs), to 670 cfs. The steel siphons are plagued with slope stability problems and leaks, and the concrete in the hydraulic drop structures is severely deteriorating. Landslides along the canal and deteriorated condition of the structures make the canal unreliable as a water source.

Failure of a hydraulic drop structure in 2002 resulted in the canal being shut down for approximately 2 months during the irrigation season. On July 21, 2004, Reclamation was forced to shut down the canal for a week to repair a leak in the left barrel of the St. Mary River Siphon. The leak was attributed to a failed weld associated with repair work completed in 2001. Within 3 days of starting water deliveries in March 2005, Reclamation was once again forced to shut down the system for another repair on the left siphon barrel.

Operation of the St. Mary Diversion & Conveyance Facilities has had a series of negative environmental impacts on the Blackfeet Reservation. Operation of the system has led to flooding and erosion below the confluence of Swiftcurrent and Boulder Creeks, and the formation of a delta into Lower St. Mary Lake. As currently designed, the outlet works on Sherburne Dam are incapable of passing low flows during the winter months. As a result, Swiftcurrent Creek dries up and important wintering habitat for the threatened bull trout (Salvelinus confluentus) is lost. In addition, the St. Mary diversion dam and canal headgates are having a negative impact on Blackfeet Tribal fishery resources. The diversion dam acts as a barrier to fish moving upstream and a large number of fish become entrained in the canal through the headgates during the irrigation season.

LIFELINE OF THE HI-LINE

The St. Mary Diversion & Conveyance Works are the foundation of large-scale irrigated agricultural production in Montana's Milk River Basin. The system provides water to irrigate over 110,000 acres through contracts with Reclamation. An additional 5,000 acres within the Ft. Belknap Indian Irrigation Project (BIA) and 25,000 acres of private irrigated land benefit from the supplemental flows. Together these farms produce approximately 8.7% of all cattle/calves produced in the state, approximately 8.2% of all irrigated hay, and 8.6% of all irrigated alfalfa produced in Montana.

In a normal irrigation season (May through September), approximately 70 % of Milk River flow near Havre originates from the St. Mary River Basin. In dry years the imported water may account for up to 90 % of the Milk River flows past Havre. During the drought of 2001, 95 % of available water in the Milk River originated in the St. Mary River Basin!

Although the St. Mary Diversion & Conveyance Works were initially built to provide supplemental water for irrigated agriculture, the beneficiaries extend far beyond this original intent. Reclamation's 2005 "Current Use Benefits Analysis" showed large public benefits accrue from the existence of the St. Mary Diversion & Conveyance Works. A preliminary economic study commissioned by the State and conducted in consultation with Reclamation indicates that 32% of the annual economic benefit associated with the supplemental water supplied by the system accrues to irrigated agriculture. The remaining 68% accrues to the public in the form of municipal, residential and industrial water, recreation, fish and wildlife, and extensive riparian areas. Over the last 90 years beneficiaries of water imported from the St. Mary River have expanded to include the following.

- Reclamation has contracts to deliver 4,664 acre-feet per year of municipal and industrial (M&I) water to the entities shown in Table 2.

The communities of Havre, Chinook and Harlem serve approximately 14,000 customers

Table 2. Municipal and Industrial Water Contracts Associated with Milk River Project

Entity	Contract Amount (Acre-Feet)
City of Chinook	700
City of Harlem	500
City of Havre	2,800
Hill County Water Users	500
North Havre County Water District	100
Grand View Cemetery	14
Horizon Hills Golf Course LLC	50

- Bowdoin National Wildlife Refuge located 7 miles east of Malta, contracts with Reclamation for approximately 3,500 acre-feet per year of supplemental water from the St. Mary River. This 15,550-acre refuge provides food and habitat for an estimated 100,000 waterfowl each spring and fall.
- Fresno and Nelson Reservoirs were created as storage components within the Milk River Project. Today these reservoirs support tremendous tourism and public year-round recreational benefits including boating, camping, and fishing. According to a 2002 Reclamation study, the Milk River Project provides approximately $15 million per year in recreational benefits to the Milk River Basin.
- The Fort Belknap Water Rights Compact is predicated on the continued viability of the St. Mary Diversion & Conveyance Works to deliver water to the Milk River Basin. The Compact is a delicate negotiated balance of water rights, including the Gros Ventre and Assiniboine Tribes' right to essentially all of the natural flow of the Milk River, subject to the claims of the Blackfeet Nation.
- The State of Montana and Blackfeet Tribe have reached agreement on a water rights compact which includes claims for water from the St. Mary and Milk Rivers. The parties are seeking federal and state ratification of the settlement. Rehabilitation of the St. Mary Diversion & Conveyance Works affords potential benefits to the settlement.
- Implementation of the 1909 Boundary Waters Treaty between the United States and Canada is dependent on the existence of the St.

Mary Diversion & Conveyance Works. Without the system, the United States' share of water from the St. Mary River under the 1909 Treaty would be lost to Canada.

- Numerous endangered, threatened, and proposed species with the Milk River Basin, including the Piping Plover (threatened) and Pallid Sturgeon (endangered), benefit from augmented flows in the Milk River.
- Water imported to the Milk River Basin through the St. Mary Diversion & Conveyance Works may prove to be a critical component of U.S. Fish and Wildlife Service recovery efforts for the endangered pallid sturgeon *(Scaphirhyncus albus)*. As the first major tributary entering the Missouri River below Ft. Peck Dam, the turbid water of the Milk River may provide critical pallid sturgeon spawning habitat.

Working with Reclamation

For many years, Milk River irrigators, State of Montana water resource staff, and Reclamation staff have been working together on water management issues in the Milk River. It was not until the mid-1990 that the potential catastrophic failure of the St. Mary Diversion Facilities was recognized by the State and irrigators as a significant risk. At the time, the Chinook Irrigation Division had received a grant of $300,000 to improve water use efficiency. However, an emergency repair of the St. Mary Siphons arose and the Chinook Division requested that the State funds be used to pay for the emergency repair rather than on water use efficiency. The funds were transferred and an additional $100,000 was added so that the siphons could be repaired and returned to use. The following year the same scenario played out where districts requested emergency funding for a repair of the St. Mary Siphon. This time the State met with irrigators and Reclamation on the St. Mary diversion site to discuss the reasons that the emergency arose. It was at this on-site meeting that concern over the potential catastrophic failure of the system was recognized.

The State then began to work with Milk River irrigators and Reclamation to find solutions so that a major rehabilitation project could proceed. Through the Rocky Boy water compact (P.L.106-163), the Montana Congressional Delegation was able to gain support for a $3 million appropriation for Reclamation to conduct the North Central Montana Regional Feasibility Study (2004). As part of the study, Reclamation examined 18 alternatives to resolve water supply, water use and management, and other major water-related issues in north central Montana. Reclamation identified rehabilitation of the St. Mary Diversion & Conveyance Works as the only feasible alternative that would address water supply and related issues in north-central Montana, and produce positive economic benefits.

Reclamation's conclusion supported State and local efforts to find a way to rehabilitate the St. Mary Diversion Facilities. However, Reclamation rather than taking a leadership position to rehabilitate the system, stated publicly that they would not support a rehabilitation plan if it involved the use of federal funds. The only acceptable approach to the agency was for the irrigators to pay for the entire cost. During the initial public meetings announcing the new Water 2025 initiative, Commissioner of Reclamation, John Keys made it clear to Montana representatives that Reclamation could not support a rehabilitation effort that involved federal funding.

Protecting the Milk River Basin's Tomorrow.......Today

The dilemma of knowing that the entire water supply of the Milk River basin was at risk and the only option that Reclamation would support would be impossible for the irrigators to support forced the State of Montana to act. On November 18, 2003, former Lt. Governor Karl Ohs held a forum in Havre, Montana, on the need to rehabilitate the St. Mary Diversion & Conveyance Works. An outcome of this meeting was the formation of the St. Mary Rehabilitation Working Group. This 15-member Working Group represents a broad coalition of basin interests including the Milk River Irrigation Districts, the Blackfeet Tribe, the Tribes of the Ft. Belknap Indian Community, municipalities, business interests, and recreational and fishery interests in the Milk River Basin. The Working Group's goals are:

1. Find a "workable" solution for rehabilitating the St. Mary Works before the system suffers catastrophic failure.
2. Work with the Blackfeet Tribe to address environmental impacts associated with the operation of the St. Mary Works, and provide workable enhancements and mutual benefits from a rehabilitated St. Mary Canal.

3. Explore options for restoring Fresno Reservoir to its original capacity and reauthorization and funding opportunities to rehabilitate the Basin infrastructure.

Though Reclamation attends all Working Group meetings, the agency has chosen to take a role limited to monitoring the discussion and providing technical information. This limited role is confusing to basin residents who do not understand why the agency that owns and operates the facilities is not at the front leading the discussion. Instead, the State of Montana has had to take the leadership role to move this project forward. Not only is it confusing for the owner/operator not to be fully engaged in the process, it is detrimental to the overall effort.

Reclamation appears to be paralyzed in the face of the tidal wave of aging infrastructure issues they face across the West. It appears that without direction from Congress, the Department of Interior will not allow Reclamation to actively participate with project irrigators and the State of Montana to rehabilitate the St. Mary system.

Through inaction at the federal level, the federal government is missing an opportunity to work with stakeholders and the State of Montana to find a workable solution for ensuring the continued viability of the one of the Nation's first reclamation projects. This federal inertia may also impact the federal government's ability to settle federal reserved water rights claims with the Blackfeet Tribe and Ft. Belknap Indian Community and jeopardize the ability of the United States to access water from the St. Mary River under terms of the 1909 Boundary Waters Treaty.

Local and State Support

Over the last four years, the State of Montana and St. Mary Rehabilitation Working Group have raised substantial non-federal funds to initiate the rehabilitation effort. Since 2004, the State has spent over $1.6 million on professional engineering services and personnel services in support of the project. The 2005 State Legislature recognized the importance of this project to Montana by approving $10 million in bonding authority to serve as seed money for the non-federal cost share. In 2007, the State Legislature reconfirmed their commitment to the project by approving a $300,000 grant to the Department of Natural Resources and Conservation for engineering design and project coordination.

Local contributions have raised approximately $416,632 to support the efforts of the St. Mary Rehabilitation Working Group (as of 12/07). The eight Irrigation Districts within the Milk River Project have assessed themselves to support the rehabilitation effort. Counties, communities, recreationists, and private citizens within the Milk River Basin have also raised funds. Since 2004, Working Group members have donated more than $156,000 (as of 10/07) of in-kind support in the form of time and travel to attend meetings and promote the project within the basin.

Federal Support

The Working Group's efforts have also benefited from support provided by Montana's Congressional Delegation. Senator Max Baucus secured $8.5 million (FY06) from the Senate Transportation Committee. These funds are being used to construct a new bridge across the St. Mary River and address environmental concerns of the Blackfeet Tribe. Senator Conrad Burns secured $500,000 (FY06) for the project from the Energy and Water Appropriations Subcommittee of the Senate Appropriations Committee. Through the use of cooperative agreements, the U.S. Bureau of Reclamation, Blackfeet Tribe, and Montana DNRC used these funds on technical and environmental data collection efforts in preparation for National Environmental Policy Act compliance activities. The Bureau of Reclamation received an additional $500,000 in FY07 under the continuing funding resolution. Reclamation has used these funds to enter into an agreement with the Bureau of Land Management to conduct a cadastral survey along the canal route. This effort is part of a multiphase endeavor to settle outstanding right-of-way issues.

Potential Solutions

The State of Montana and St. Mary Rehabilitation Working Group have put in a tremendous effort to raise non-federal funds to initiate the rehabilitation project. To date, State and local funds committed to rehabilitation of the St. Mary Diversion and Conveyance Works exceed $12.5 million. What is missing at this point is active participation by Reclamation to find a workable solution. Even with the grass roots support of the basin and the leadership and financial support of the State, we cannot succeed without the support of Congress and, in turn, Reclamation.

Bureau of Reclamation must be given the ability to develop new tools for addressing the challenges aging water resources infrastructure. One challenge we face in the Milk River Basin is how the Bureau of Reclamation views capital expenditures for rehabilitation projects. According to a 2006 engineering study undertaken by the state of Montana the cost to rehabilitate and replace the St. Mary Diversion & Conveyance Works is estimated at $130 million to $140 million. The Bureau Reclamation considers such work to be part of operation and maintenance and requires the cost to be paid back in the year they are incurred. This presents an insurmountable financial obstacle to irrigators in the Milk River Project. Irrigators within our Project would benefit from the development of a program that enables water users to modernize or rehabilitate their projects and payoff those costs over time under reasonable terms and conditions.

As originally authorized, the St. Mary Diversion & Conveyance Works are operated for the single purpose of irrigation. As such, nearly 100% of the cost to rehabilitate and replace the system must be borne by small irrigators like me, within the irrigation districts holding water delivery contracts. Today the beneficiaries extend far beyond this original intent. The Bureau Reclamation's 2005 "Current Use Benefits Analysis" showed large public benefits accrue from the existence of the St. Mary system. In addition to providing water for irrigated agriculture, the system provides water to communities and industry, water to Bowdoin National Wildlife Refuge, and water that supports a wealth of recreational opportunities in north-central Montana. The system is also integral to settlement of federal reserved water rights with the Blackfeet Tribe and Ft. Belknap Indian Community, and implementation of the 1909 Boundary Waters Treaty. Amending the original authorization for the Milk River Project and St. Mary Diversion and Conveyance Works to reflect the extensive public benefits they provide would relieve irrigators of the burden of subsidizing the federal government for the benefits enjoyed by others.

The Working Group and the State are eager and willing to work with Reclamation and Congress to find acceptable solutions for federal support to the project. We understand that National priorities make it extremely difficult to find federal funds for the rehabilitation of the St. Mary Diversion Facilities. The tremendous amount of authorized, but not completed Reclamation projects also places a heavy burden on the agency. It is important to remember that many of these Reclamation projects are federal assets owned and operated by the federal government. Investment in these projects will preserve our Nation's ability to conserve, enhance, and efficiently manage our most precious natural resource, WATER! There is not time to simply wait for the federal funding environment to change.

In: Water Infrastructure
Editor: Harald Nystroem
ISBN: 978-1-60692-479-2

Chapter 4

TESTIMONY OF TONY WILLARDSON, WESTERN STATES WATER COUNCIL ON BEHALF OF THE WESTERN GOVERNORS' ASSOCIATION AND THE WESTERN STATES WATER COUNCIL BEFORE THE SUBCOMMITTEE ON WATER AND POWER OF THE ENERGY AND NATURAL RESOURCES COMMITTEE UNITED STATES SENATE

Tony Willardson

Dear Mr. Chairman and Members of the Subcommittee:

My name is Tony Willardson. I am Deputy Director of the Western States Water Council, and a member of the Western States Water Council, an organization of eighteen western states. Our members are appointed by their respective governors, and include senior state water managers and administrators. We are also closely associated with the Western Governors' Association (WGA), and I am testifying on their behalf. Water resources infrastructure investments and financing those investments are an important topic for the Governors and the Council. In April 2005, the Council addressed the full Committee as part of a Water Conference on various related topics, including the future role of the Bureau of Reclamation.

Our June 2006 report, *Water Needs and Strategies for a Sustainable Future*, highlighted support for two infrastructure related items within the jurisdiction of this Committee: (1) the Rural Water Supply Act since enacted into law; and (2) appropriations from annual receipts "...accruing to the Reclamation Fund for authorized Bureau of Reclamation projects and purposes, to help meet western water supply needs, especially for rural communities, to maintain and replace past projects, and to build new capacity necessary to meet demands related to growth and environmental protection." We look forward to working with Reclamation under the Rural Water Supply Act to assess related infrastructure needs. I will say more about the Reclamation Fund later.

The Water Report also directed the Council to hold a series of symposia designed to "(a) bring stakeholders together to try and find ways to meet our growing western water, wastewater, watershed protection and restoration, and public safety-related infrastructure funding needs; (b) find ways to quantify, evaluate and prioritize funding those needs; and (c) highlight the benefits of integrated watershed, riverbasin, regional and interstate planning and management."

Last month, the WGA and WSWC joined the Interstate Council on Water Policy and the Association of State and Interstate Water Pollution Control Administrators to cosponsor a "Water Infrastructure Workshop." To the extent the findings and recommendations from that discussion were consistent with our past positions and testimony, they have been incorporated into today's testimony. One recommendation which we whole-heartedly endorse is the need for the President and the Congress to provide adequate resources for the collection of basic water data to provide the science necessary to evaluate present and future water infrastructure needs. Last December, the Council testified in support of the SECURE Water Act before the full Committee. Other workshop suggestions will be considered, but have yet to be vetted by the WSWC and approved by the Governors.

Various estimates of our Nation's total infrastructure needs have been made. The American Society of Civil Engineers (ASCE) testified before the Senate Banking Committee last month that "aging and overburdened infrastructure threatens the economy and quality of life in every state, city, and town in the nation." ASCE's *2005 report card for America's Infrastructure* presented an overall grade of "D" and ASCE estimates it would take "an investment of $1.6 *trillion* by 2010 to bring the nation's existing infrastructure into good working order."

The Bureau of Reclamation operates hundreds of dams and reservoirs in the West supplying water and power to millions of people, irrigating millions of acres for food and fiber, providing flood control and recreation, and maintaining instream flows for fish and wildlife habitat, including anadromous and threatened and endangered aquatic species. The value of federal Reclamation projects can not be overstated, particularly in assisting western communities to endure extended drought that continues to afflict parts of the West. Two of Reclamation's express "mission goals" are: (1) managing, developing and protecting water and related resources to meet the needs of current and future generations; and (2) operating and maintaining facilities safely, reliably, and efficiently to protect the public investment.

These investments are key to our Nation's continued economic prosperity and further environmental improvements. The West faces a continuing need for storage in Reclamation projects, and in the future more storage will need to be considered as part of an integrated water resources strategy that relies on a mix of water supplies. It is essential that we preserve and protect our existing water infrastructure, if we are to maintain past gains in environmental quality and our present and future quality of life. In general, we are not yet in a crisis, but face a chronic problem that will only get worse without aggressive, affirmative action. If we are to leave a firm water infrastructure foundation for future generations, we will need to increase spending for project repairs, replacement and new construction. Their water future is in our hands. Our decisions, actions or inaction, will affect not only our quality of life, but theirs. Similarly, we enjoy the foresight of earlier generations.

Reclamation faces many serious challenges to balance and provide for a new mix of resource needs in the West due to population growth and changing values. While traditional agricultural demands continue to dominate water use in the West, environmental uses have become more important to the public, while municipal and industrial development is demanding more and more high quality water. Climate uncertainty increases the challenge. In the future, there will be even greater demands placed on the West's limited water resources and Reclamation's aging projects, many of which are well beyond their designed life. We must preserve our existing storage capacity, and consider additional water storage alternatives, again as part of an integrated water resources planning strategy.

In the West, we all have an important interest in federal Reclamation project rehabilitation, and the solutions to problems related to aging infrastructure will require the formation of true partnerships In the foreseeable future, Reclamation faces the need to spend hundreds of millions of dollars for general operation and maintenance expenses, extraordinary maintenance, dam safety, project rehabilitation and betterment, and water delivery system improvements. The Council and the Governors recognize the need to work together as federal-state-local partners to address this challenge. Reclamation's mission goals cover a number of long-term goals that include ensuring effective operations of facilities; and operating, maintaining and rehabilitating facilities to ensure reliability and cost-effectiveness. The Bureau's strategy for accomplishing these goals lists several guiding principles that involve a continuing and close working relationship with traditional water users, while using partnerships to create sustainable solutions and leverage limited resources. Transparent decisionmaking is one important principle that can not be over emphasized. States and local districts want to be "partners" not just "paying customers." They need to be actively involved in the evaluation, design and selection of alternatives.

Moreover, project modifications, reoperations and reauthorization should be considered, as necessary and appropriate, to look at current water problems and opportunities to increase project water yields to make more water available for new and expanded uses and increase water use efficiency. Reclamation facilities also play a key role in storing, managing and delivering water under numerous interstate compacts and international treaties and tribal water right settlements. These facilities must be maintained and operated so as to insure that U.S. interests are protected, and U.S. obligations fulfilled.

The Administration and the Congress must make this a budget priority. How should Reclamation programs and projects be funded? Created by the Reclamation Act of 1902, the Reclamation Fund was envisioned as the means to finance western water and power projects with revenues from western resources. Its receipts are derived from water and power sales, project repayments, certain receipts from public land sales, leases and rentals in the 17 western states, as well as certain oil and mineral-related royalties. It is a special fund within the U.S. Treasury that is only available for expenditure pursuant to annual appropriation acts.

With growing receipts, in part due to high energy prices, and declining federal expenditures for Reclamation purposes, the unobligated balance gets larger and larger (and otherwise reduces the need for federal borrowing proportionally), with the money actually spent elsewhere for other purposes. Receipts in the past were insufficient for the construction of major federal projects such as Grand Coulee and Hoover Dams, which required the appropriation of general Treasury funds, but today it appears that the

Reclamation Fund balance is more than sufficient to pay for Reclamation's water resources programs at current levels.

The Congress, concerned with budget scoring problems, rejected the Administration's FY 2006 request to allow Reclamation to spend certain revenues from water and power receipts in the Reclamation Fund for project operation and maintenance expenses without further appropriation. At present, Reclamation Fund receipts, including energy-related revenue from federal lands, exceed appropriations by roughly $1 billion annually. The WGA and the Council strongly believe the Administration should request and the Congress should appropriate more of this money for Reclamation project operation, maintenance, rehabilitation and replacements -- as well as to "...build new capacity necessary to meet demands related to growth and environmental protection," in close consultation with western states. (Water Report 3B, p. 15)

The President's FY 2009 budget request for Reclamation's Water and Related Resources account totals just over $779 million, compared to actual FY 2008 appropriations of over $949 million -- continuing a general downward trend. Current program and financing figures and estimates of new budgetary authority (gross) for obligation for FY 2009 is $1.053 billion, down from $1.113 billion for FY 2008 and $1.074 billion in FY 2007. Total gross outlays would be $1.077 billion, compared to an estimated $1.435 billion in FY 2008 and just over $1 billion in FY 2007. Reclamation's facility maintenance and rehabilitation figure for FY 2009 is $183 million, compared to $195 million for FY 2008, and $201 million for FY 2007. Obviously, spending on Reclamation infrastructure is going in the wrong direction.

Meanwhile, the unobligated balance in the Reclamation Fund continues to grow. The actual balance at the end of FY 2007 was $6.567 billion – and the estimated balance at the end of FY 2008 is $7.612 billion – with an estimated balance at the end of FY 2009 of $9.232 billion. (By way of comparison, the Administration estimated that the fund at the end of FY 2006 would be $5.905 billion, but the actual balance was $5.671 billion). The actual balance at the end of FY 2004 was $3.877 billion. Between the end of FY 2004 and the end of FY 2009, the unobligated balance will have grown by $5.36 billion, if current FY 2009 estimates are correct.

Next, federal loan guarantees authorized by Congress should allow the Bureau of Reclamation to provide the means for the repayment of state and local bonds for the rehabilitation and construction of projects. The WSWC has in the past also supported a similar insurance fund, as well as the use of tax-exempt bonds to finance water resources needs. Tax credit bonds are another potential tool. State and local agencies finance the majority of their own water needs, but federal assistance has and will continue to be important.

Interior's Water 2025 Initiative is an example of Reclamation's efforts to address water resources challenges in the West before conflicts reach a critical impasse. The success in leveraging federal, state and local resources through its Challenge Grants is an example of what can be accomplished if we are willing to work together. It is apparent that matching non-federal support could easily be found for $100 million in federal money, although this program alone is insufficient to provide meaningful support for water infrastructure needs in the western United States.

It is time to focus federal financial resources intended to aid in western water development to help Reclamation and state and local agencies meet the future challenges of continuing to supply adequate water of suitable quality in the face of growing municipal and industrial demands and federal requirements to protect public health and the environment. Federal infrastructure investments are justified in order to maintain our Nation's economic and environmental vitality, to assist state and local entities meet federally mandated standards, and to aid economically stressed communities.

Water has always been a valuable commodity in the West, and it is now increasingly so. We should all expect to pay more for water and water-related goods and services in the future as individuals and as governments to invest more in our water resources infrastructure. Regional projects offer potential savings due to economies of scale, but "exburbanization" including the development of whole subdivisions in rural areas with each homeowner relying on their own well and septic system are creating new stresses. Federal, state and local water and land management and planning agencies need to work together on a watershed and river basin scale to find solutions. More research is needed into water supply alternatives, asset management tools, risk assessment and acceptable risks, etc.

Current challenges may provide an opportunity to look beyond existing ownership and partnership arrangements, as well as authorized project purposes and benefits. Appropriate public-private partnerships should be considered where they have the potential to accelerate high priority projects, fast track financing or provide incentives to maximize performance. However, such contracts require clearly defined responsibilities and performance requirements.

In the future, as we address the growing water needs in the West for many purposes, different agencies and stakeholders may have to come together and pool their available financial and other resources in new project specific partnerships, as well as nonstructural agreements, in order to overcome the challenges and obstacles we face in resolving our aging infrastructure problems and insuring the West and the Nation has an adequate water supply.

On behalf of the Western Governors' Association and Western States Water Council, our members and member states, we appreciate the opportunity to testify and I would be happy to answer any questions. Thank you.

In: Water Infrastructure
Editor: Harald Nystroem
ISBN: 978-1-60692-479-2

Chapter 5

TESTIMONY BEFORE THE UNITED STATES SENATE ENERGY AND NATURAL RESOURCES COMMITTEE'S, SUBCOMMITTEE ON WATER AND POWER, AN OVERSIGHT HEARING ON AGING INFRASTRUCTURE

Thomas F. Donnelly

Mr. Chairman, members of the Committee, my name is Thomas F. Donnelly and I am the Executive Vice President of the National Water Resources Association. On behalf of the membership of the Association, it is my privilege to present testimony on the issue of the Bureau of Reclamation's aging infrastructure.

The National Water Resources Association (NWRA) is a nonprofit federation of associations and individuals dedicated to the conservation, enhancement, and efficient management of our Nation's most precious natural resource, WATER. The NWRA is the oldest and most active national association concerned with water resources policy and development. Its strength is a reflection of the tremendous "grassroots" participation it has generated on virtually every national issue affecting western water conservation, management, and development.

In the next decade, we believe, that one of the most critical problems facing the Bureau of Reclamation and many water districts throughout the West is maintaining the existing water and power infrastructure at peak operational efficiency. Many projects have met or exceeded their design life and are in need of modernization and/or rehabilitation. This is not to say that these projects cannot efficiently provide benefits well into the foreseeable future. However, modernization and rehabilitation of these important facilities will be required.

Currently, the U. S. Bureau of Reclamation (USBR) does not have a program which enables water users to modernize or rehabilitate their projects and payoff those costs over time under reasonable terms and conditions. Such works are considered operation and maintenance and consequently the costs must be paid back in the year that they occur. The Bureau understands the problem but has not been able to develop a program to address the problem that is acceptable to the Office of Management and Budget. This is a problem that, if not addressed immediately, will result in severe consequences sooner rather than later.

Neither the Bureau of Reclamation nor the membership of the National Water Resources Association has an accurate grasp of the scope of the problem West-wide. Therefore, we recommend that the first step in addressing this issue should be an accurate assessment of the scope of the problem.

From a funding and repayment perspective, the projects can essentially be divided into three rather distinct categories; 1.) those multi-purpose projects with a vendible output such as power or municipal water supply whose operating revenues adequately allow for future rehabilitation and modernization, 2.) those multi-purpose projects with operating revenues sufficient to repay the cost of major rehabilitation and modernization over time given reasonable terms and conditions, and 3.) those single purpose irrigation projects whose initial construction was heavily underwritten by the federal government and whose beneficiaries do not possess the financial resources to repay the costs of modernization and rehabilitation.

For the purposes of this hearing we need only to consider those projects falling into category 2 and 3.

PROJECTS WITH REPAYMENT CAPACITY

For the second category of projects (with repayment capacity over time), the solution is simply a program of funding that allows the beneficiaries to repay the modernization and rehabilitation cost over a reasonable period of time under fiscally sound terms and conditions. As stated earlier, the Bureau of Reclamation does not currently have a program which facilitates the timely modernization, rehabilitation or replacement of its existing infrastructure.

An example of a project that falls into this category and needs immediate assistance is the Minidoka/Burley Project in Idaho.

Over the past four years, we have engaged in discussions with Bureau's representatives concerning this problem. Unfortunately, the Bureau of Reclamation is tremendously restrained with what it can offer as a solution by the Administration's Office of Management and Budget and consequently, unable to think "outside the box." In the 109th Congress Reclamation successfully promoted loan guarantee legislation which was signed into law by President Bush. In some instances, loan guarantees could work, but loan guarantees do not offer a comprehensive solution. Ironically, now the Office of Management and Budget has cooled to the concept of loan guarantees and has stymied Reclamation's effort to promulgate regulations to implement this legislation which the President enthusiastically signed.

We believe there are a number of potential solutions, some using existing authority and others requiring new program authorization. Modernization and or rehabilitation of these projects could be accomplished through various means: project specific authorization (amendment of original authorization or new authorization) and appropriations, a congressionally authorized U. S. Bureau of Reclamation modernization and rehabilitation program (PL 81-335), infrastructure revolving fund or use of the Reclamation Fund. A restructuring of the Reclamation Fund, established under Section 1 of the Reclamation Act of 1902 (32 Stat.388; 43 U.S.C. §391), is an example of a potential solution using existing authority. The "Fund" currently has approximately $6-7.0 billion in it. In reality, it was envisioned to address both new project construction and the modernization and rehabilitation of the existing infrastructure.

PROJECTS WITHOUT REPAYMENT CAPACITY

The second category of projects poses a much more difficult set of public and fiscal policy questions for Congress and the Administration.

Many of the early projects authorized and built under the Reclamation Act were single purpose irrigation projects. They were built in an effort to develop regional agricultural economies in order to facilitate a national policy goal of the orderly development of Western lands. Regional economies have developed around these projects. In many instances, project benefits have been expanded to include municipal and rural water supply, environmental mitigation and Native American water rights settlements. Despite the economic development, project beneficiaries of these single-purpose projects do not possess the repayment capacity to repay the cost of major modernization or rehabilitation.

The St. Mary Diversion project in north central Montana is a classic example of this second category of projects. The project was authorized and built under the Reclamation Act of 1902 to provide a stable source of water for irrigation of the lower Milk River Valley and settlers moved to the valley on that promise.

The USBR currently estimates that rehabilitation of the St. Mary Diversion facilities will cost approximately $130 million. The State of Montana and the local beneficiaries would be pressed to pay even a quarter of the estimated costs. While the State of Montana has committed $10 million toward the cost of rehabilitating the project, it is hard to imagine why the state would invest much more in a purely federal asset. Therefore, the necessary rehabilitation costs will require a substantial federal re-financing of the project.

Conversely, the cost of letting the project further deteriorate would be enormous. A sudden failure would result in enormous environmental damage on the Blackfoot Indian Reservation, economically devastate local communities and businesses and undermine the Fort Belknap Water Rights Compact. It is likely that the State of Montana and the agricultural economy in the northern tier states would also be adversely impacted.

It is important not to lose sight of the fact that these projects are still federally owned facilities with the underlying responsibility of the federal government to operate and maintain them at peak efficiency.

SUMMARY

The water supply and power infrastructure built over the last century by the Bureau of Reclamation remains vitally important to the West and the Nation as a whole. Reclamation projects authorized by Congress provide numerous and substantial benefits for the entire United States. Among these benefits are: (1) flood prevention and protection totaling in the tens of billions of dollars; (2) generation of substantial amounts of hydroelectric energy using water as a renewable no-cost fuel source; (3) delivery of irrigation water to hundreds of thousands of acres of farmland in semiarid and arid regions that has increased and stabilized agricultural production in those regions; (4) water-based outdoor recreation facilities that provide recreation for millions of visitors annually; (5) municipal and rural domestic water supplies for over 30 million people; (6) recharge of underground aquifers and water supplies; (7) fish and wildlife habitat including new fisheries, wildlife management areas, and hundreds of thousands of acres of habitat and marshes throughout project distribution systems and facilities; and (8) major surface water transportation.

We simply cannot let this amazing legacy crumble and deteriorate. The membership of the National Water Resources Association pledges its support and assistance to the Committee as it seeks solutions to this important problem. In addition, we commend the Chairman and the Committee for their recognition of the aging infrastructure problem and timely attention to finding a solution.

Respectfully submitted,

Thomas F. Donnelly
Executive Vice President
National Water Resources Association
3800 North Fairfax Drive, Suite #4
Arlington, VA 22203
(703) 524-1544
tdonnelly@nwra.org

In: Water Infrastructure
Editor: Harald Nystroem
ISBN: 978-1-60692-479-2

Chapter 6

Statement of Robert W. Johnson, Bureau of Reclamation, U.S. Department of the Interior, before the Committee on Energy and Natural Resources, Subcommittee on Water and Power U.S. Senate, Reclamation's Aging Infrastructure

Robert W. Johnson

Mr. Chairman and members of the subcommittee, I am Robert W. Johnson, Commissioner of the Bureau of Reclamation. I am pleased to provide the Department of the Interior's perspective on Reclamation's asset management strategy as it relates to our aging water and power infrastructure. I will also discuss some of the reinvestment needs we have identified.

Before I start, I would like to note that I have been a Reclamation employee for over 30 years, and am well aware of the challenges we face as a result of our aging infrastructure. Operating water facilities, whether aging or newly complete, brings with it many unknowns and engineering challenges. A recent example of these challenges is provided in the case of the Truckee Main Canal (TMC), near Fernley, Nevada. Early on the morning of January 5, 2008, a portion of the Truckee Canal embankment failed resulting in uncontrolled water releases into residential areas of the City, causing varying damage to

590 homes. The canal, operated and maintained by the Truckee-Carson Irrigation District under a contract with Reclamation, provides water to agricultural and wetland uses in the Fernley and Fallon, Nevada areas. At the location of the breach, the canal, built a hundred years ago, has earthen embankments and is unlined.

After the breach, the District shut down the canal and placed a temporary earthen plug into the breach site to stem flows into the City. Concurrently with designing a permanent repair for the breach, Reclamation initiated several studies and investigations with the purpose of determining likely contributing factors to the failure, the condition of the remainder of the 31-mile long canal, the risks associated with resuming flows in the canal through the reach above the City, and the conditions under which deliveries might be resumed. The inspection of the remainder of the canal was conducted by teams, including members from the City, the District and the Corps of Engineers. A team of independent experts determined that the most likely contributing factor to the failure was rodent activity.

During March, staged diversions into the canal were resumed beginning at 20% of the maximum canal flow with ramping allowed up to 45% of the maximum flow. Flows above 20% of the maximum are authorized only if the District meets specific requirements, such as development of emergency action, maintenance and facility improvement plans. Flows above 33% are allowed only after a special rodent control program has been carried out. Flows may not exceed approximately 45% of the maximum flow until a permanent fix is in place, such as an impermeable barrier on the city side of the canal protecting the City of Fernley. The staged flow restrictions are the direct result of the studies and investigations undertaken by Reclamation – designed to determine under what circumstances diversions could be resumed under safe and reliable conditions.

It is important to recognize the extent of the damage caused by the failure, but it is also important to recognize that Reclamation was able to assemble the expertise necessary to evaluate the conditions under which flows might safely be resumed and develop a flow regime under which the District could resume restricted diversions within eleven weeks. It took an intensive, concerted effort by Reclamation and the District to achieve this result. The Reclamation contribution involved staff from the area and regional offices, together with staff in Reclamation's Technical Service Center.

The TMC is over 100 years old. Given that Reclamation's first projects were started over 100 years ago, our employees and managing partners have done an incredible job of safely maintaining and protecting our infrastructure.

In the aftermath of events on the TMC, Reclamation has initiated a new effort to ascertain where growth has occurred around canals, and where a similar condition may exist that could pose a threat to life or property. We intend to work with our operating partners to select representative canal reaches in urbanized areas within each of our five regions for special reviews to be conducted this year. This process will afford the opportunity to engage interested operating partners in the topics of asset management and addressing our aging infrastructure.

While Reclamation's reach across the West is widespread, our employees take the safety of our facilities and the protection of our customers and surrounding communities very seriously. And, as a result, the vast majority of our infrastructure is in good working order. I am very proud of our record.

In fiscal year 2009, Reclamation's Dam Safety Program plans to have corrective actions underway at six facilities across the west, and is requesting an increase of over $15 million above the amount appropriated in fiscal 2008 in the Dam Safety Program. Reclamation has also requested funds to study the need for potential corrective actions at other facilities.

Reclamation's mission is to "manage, develop, and protect water and related resources in an environmentally and economically sound manner in the interest of the American public." We are the Nation's largest wholesale water supplier, and the 348 reservoirs we administer have a total storage capacity of 245 million acre-feet of water. We bring water to more than 31 million customers and provide 20 percent of western farmers with water to irrigate 10 million acres of farmland. Reclamation is also the Nation's second largest producer of hydroelectric power, generating more than 40 billion kilowatt-hours of energy each year – equivalent to the energy provided by 80 million barrels of crude oil. In the 100 years since Reclamation's creation, the Federal government has invested almost $21 billion in original development costs for our facilities. The current cost to replace these assets would be many times that original development cost.

Reclamation's core mission has remained constant since its inception, but the way we accomplish that mission has evolved. Today, we focus primarily on managing and maintaining our facilities to ensure their safe and effective operation while continuing to deliver water and power. In terms of actual operation and maintenance, Reclamation operates about one-third of its facilities, and the other two-thirds (primarily single-purpose irrigation facilities) are operated and maintained by non-Federal operating entities (e.g., water/power districts formed under state laws to provide service to a particular area or set of customers).

Most of Reclamation's major dams, reservoirs, hydroelectric plants, and irrigation systems are 50 or more years old. A central point we would like to make is that a facility's age by itself is not the sole determinant of its viability – rather, facility condition is the central factor in predicting the long-term functionality and maintenance need of Reclamation assets. As part of Reclamation's oversight of constructed assets, we initiated a Facility Review Program in 1948 to assess the condition of assets constructed by Reclamation and operated and maintained by our non-Federal operating partners. These activities continue today and, in concert with a preventive maintenance philosophy and related oversight initiatives, have successfully extended the service life of many of our water and power facilities beyond original expectations. Reclamation has recently been taking steps to more accurately represent its inventory of assets in the Federal Real Property Profile (FRPP).

Nevertheless, the aging of our infrastructure constantly presents new maintenance, replacement, and modification requirements. Similar to other agencies with aging infrastructure, Reclamation has a fiduciary duty to maintain services to its customers in a cost efficient manner and to meet other expectations, particularly environmental and endangered species management. While Reclamation and over 350 operating partners have for many years operated and maintained the infrastructure, the very nature of the aging process will inevitably lead to increased pressure on budgets and user rates to keep infrastructure service and reliability commensurate with past levels. As such, Reclamation and the operating entities anticipate a steady increase in infrastructure repair needs that will continue to grow over time. As part of Reclamation's asset management strategy, regular operation and maintenance activities under appropriated dollars will be managed in concert with other programs and activities addressed in our strategy to improve efficiency and effectiveness in funding rehabilitation and replacement needs.

Improved technology will also offset many of these costs, as will innovative construction processes like the one occurring on the Joint Federal Project at Folsom Dam near Sacramento. Together with the U.S. Army Corps of Engineers, Reclamation is undertaking an historic effort to jointly construct features that will address both safety of dams concerns, as well as expand flood protection for the City of Sacramento. Separately, these two projects would cost over $2 billion and would take 15 years to complete, but by working together to design and construct features consistent with these two distinct activities, Reclamation and the Corps estimate that the joint project should cost half that much and be completed in half the time. Project construction is planned to proceed in phases by Reclamation and the Corps.

Procedurally, Reclamation's Facility Maintenance and Rehabilitation Program identifies, schedules and prioritizes necessary rehabilitation work at reserved works. To fulfill these responsibilities, Reclamation provides designs and studies, purchases equipment and services, and provides the resources to support the overall maintenance and rehabilitation program. Project beneficiaries advance funds for annual O&M work performed by Reclamation.

However, for many other facilities, rehabilitation and replacement needs may exceed available resources and could potentially increase the risk of service interruption or failure, as occurred early this year on the Truckee Main Canal. To fund this work, in cases where operating partners cover a portion of the O&M costs for reserved works, or the entire O&M costs on transferred works, the use of the entity's reserve fund is one of the first places we look for funding. However, these funds may not be contractually required nor sufficient to meet the amount needed for major rehabilitation and replacement work. Thus, long-term financing must often be obtained to fund such work, and arrangements are made with operating entities depending on the circumstances of a given project.

West-wide, Reclamation currently estimates that approximately $3 billion will be required to rehabilitate, replace, and modify Reclamation assets under major rehabilitation and replacement programs in the future. This figure was derived from a very rough, field-level estimate of conceivable needs, and includes work under our Safety of Dams program, anticipated work on our water and power reserved works, and preliminary estimates for transferred works operated by our customers. From a programmatic perspective, much of this data is insufficiently reliable to serve as a basis for budgeting or long-term planning decisions. A substantial part of projected needs will be financed directly by our water and power customers and the sale of hydroelectricity. Some funds may need to come from appropriations, but the magnitude and timing of such funding needs is unknown. As noted above, those entities which contractually operate transferred works are also required to fund operation, maintenance, rehabilitation, and replacement work at their cost, and this amounts to at least $800 million of the $3 billion estimate.

One of the challenges we face is the varying economic strength of our operating partners. For some of these partners, the cost-share requirements associated with the review and repair activities are simply beyond the means of the beneficiaries. The Administration has and will continue to be opposed to projects that are authorized without adequate cost controls and built-in accountabilities to ensure that the Federal government is not subject to undue costs. While circumstances for each project in need of review, rehabilitation or repair may be different, in order for projects to be sustainable, the non-Federal sponsors must be responsible for a fair share of project costs and, for facilities that are being operated and maintained by non-Federal entities, these entities must be accountable for maintaining the assets.

Title II of the Twenty-First Century Water Works Act (P.L. 109-451) authorized loan guarantees for eligible projects. Currently, Reclamation is working on the proposed rules for implementing this program.

Sound and reliable infrastructure is the core of Reclamation's mission. With the support of Congress, our customers, and other stakeholders, Reclamation will continue to assure the integrity and reliability of Federal water and power assets.

This concludes my written statement. I would be pleased to answer any questions.

In: Water Infrastructure
Editor: Harald Nystroem
ISBN: 978-1-60692-479-2

Chapter 7

TESTIMONY OF MAYOR TODD CUTLER, CITY OF FERNLEY, NEVADA, SENATE SUBCOMMITTEE ON WATER AND POWER

Mayor Todd Cutler

Mr. Chairman, Members of the Subcommittee, thank you for the opportunity to appear before you to discuss the impacts of a failed Bureau of Reclamation facility that has dramatically impacted the community I represent. I am Todd Cutler, Mayor of the City of Fernley, Nevada.

Our City is located 32 miles east of Reno, Nevada along Interstate 80. Our booming community has also grown up next to the Truckee Canal which runs along the west and south side of Fernley. Constructed in 1906 as part of the Newlands Reclamation Project and as one of BOR's first projects, the Truckee Canal is the primary water source to the desert cities and towns of northern Nevada. These communities along with our City rely on the canal's water for irrigation and recharge of our ground water resources. However, on January 5th of this year, the canal that we and others have relied on for current and future water resources quickly turned on us.

At 4:00 am on January 5th, the northern bank of the Truckee Canal failed, sending a wall of water rushing through our community. Our families were awakened on that cold morning by these waters; it filled their homes and quickly destroying what lied in its path. Approximately 585 homes were impacted from the water that flowed from the canal, and approximately 200 homes sustained substantial damage. Many of our homes sat in water of levels up to eight feet deep for several days.

We consider ourselves fortunate to have only suffered property loss. We were very lucky, not a single life was lost during this disaster. The evacuation efforts began immediately and were quite efficient. I can't say enough about the efforts of our local citizens, our County and State emergency partners and the help that we received from FEMA. President Bush signed the disaster declaration for the Fernley area on January 8th. The people of Fernley are thankful for everyone's efforts and are resilient, hard working, and dedicated to our community. Now we still face perhaps the biggest challenge of this disaster and that is how to repair these homes, replace lost property, and fully restore our City.

Though we may not have lost life, the effected residents lost much of their livelihood. Thousands of man-hours have been expended by our citizens and volunteers from across the State, our citizens still face the daunting challenge of how to pay for the losses they suffered. While FEMA was on the ground within hours of the flood, they only have so many tools at their disposal. Their primary purpose is to get people into temporary housing and to keep them fed and safe. FEMA assistance does not repair or replace homes.

Our residents will be looking to the federal government to fully compensate them for their losses. Many law suits have already been filed against TCID, Fernley, Lyon County and others for recovery of losses due to the canal breach has been named in some of those suits. One of my biggest fears is that lawsuits will do nothing more than tear our community apart. We need the Truckee Canal to provide water; and we need TCID to continue delivering water to our community. Lawsuits will do nothing more than line the pockets of lawyers, perhaps bankrupt TCID, and take years to resolve. We are working hard with the Nevada Congressional Delegation to find a solution to these issues. Not only are we facing perhaps $55 million in damages to homes and property, we need to insure the Truckee Canal continues to supply water but not fail in the future.

This is not the first time the Truckee Canal has failed. The canal has failed eight times in its history due to both natural occurrences such as ice jams, but also due to structural failures. Given the age and track record of this federal facility, my community remains highly skeptical about the safety of the canal and many worry that the canal will fail again. As far as the investigators can tell, piping due to rodent activity is the most likely cause of the Truckee Canal failure. The millions of dollars of FEMA, State, and local funds spent on the disaster recovery in Fernley would have been better spent on maintaining the Truckee Canal.

I whole heartedly support the legislation introduced by Senate Majority Leader Reid, known as the "Aging Water Infrastructure and Maintenance Act". Had this legislation been in place prior to January 5th of this year, my community would not be facing the challenge of getting people back in their homes and we would not be knocking on the doors of Congress to pay for damages in excess of $50 million dollars. It is my belief that the Federal Government must bear some of the burden of inspecting and repairing these major facilities if they are going to maintain ownership of them. One recommendation I would have for the Subcommittee is to review the insurance requirements of the operators of these Federal facilities. Fernley needs this canal to continue to be an integral part of our community and our future. More importantly, we need to assure our residents that the safety of the canal is a priority. This safety can only be achieved by the regular inspections and maintenance of the canal.

Thank you for your time and I am happy to answer any questions you might have.

In: Water Infrastructure
Editor: Harald Nystroem

ISBN: 978-1-60692-479-2

Chapter 8

WATER INFRASTRUCTURE FINANCING: HISTORY OF EPA APPROPRIATIONS*

Claudia Copeland

SUMMARY

The principal federal program to aid municipal wastewater treatment plant construction is authorized in the Clean Water Act (CWA). Established as a grant program in 1972, it now capitalizes state loan programs. Authorizations since 1972 have totaled $65 billion, while appropriations have totaled $78.3 billion. It has represented 25-30% of total funds appropriated to the Environmental Protection Agency (EPA) in recent years.

In appropriations legislation, funding for EPA wastewater assistance is contained in the measure providing funds for the Department of the Interior, Environment, and Related Agencies, which includes EPA. Within the portion of that bill which funds EPA, wastewater treatment assistance is specified in an account now called State and Tribal Assistance Grants (STAG). Three trends in the funding of this account are most prominent: inclusion of non-infrastructure environmental grants to states, beginning in FY1993; increasing number and amount of special purpose grants since FY1989; and the addition of grant assistance for drinking water treatment projects in FY1997. This

* This is an edited, reformatted and augmented version of a CRS Report for Congress publication dated August 2008.

report summarizes, in chronological order, congressional activity to fund items in this account since 1987.

Prior to the 1987 amendments, wastewater treatment assistance was provided in the form of grants made to municipalities. The federal share of project costs was generally 55%; state and local governments were responsible for the remaining 45%. The 1987 amendments altered this arrangement by replacing the traditional grant program with one that provides federal grants to capitalize state clean water loan programs, or state revolving funds (SRFs). As a general matter, states and cities support the program changes and the shift to a loan program that was intended to provide long-term funding for water quality and wastewater construction activities. However, the change means that local communities now are responsible for 100% of projects costs, rather than 45%, because they are required to repay loans to states. The greater financial burden of the act's loan program on some cities has caused some to seek continued grant funding.

This has been particularly evident in the appropriations process where, in recent years, Congress has reserved as much as 30% of funds in the STAG account for special purpose grants directed to specified communities. Since FY2000, appropriators have awarded earmarks to a larger total number of projects, resulting in more communities receiving such grants, but at the same time receiving smaller amounts of funds, on average. Most of the funded projects are not authorized in the Clean Water Act or the Safe Drinking Water Act. State water quality officials, state infrastructure financing officials, and EPA have objected to this practice, since it reduces the amount of funding for state SRF programs. Since FY1997, the STAG account also has been used to fund a drinking water SRF grant program established by Congress in 1996. Appropriations for the drinking water SRF program through FY2008 have totaled $10.3 billion.

INTRODUCTION

The principal federal program to aid municipal wastewater treatment plant construction is authorized in the Clean Water Act (CWA). Congress established this program, essentially in its current form, in the Federal Water Pollution Control Act Amendments of 1972 (P.L. 92-500) (although prior versions of the act had authorized less ambitious grants assistance since 1956). Title II of P.L. 92-500 authorized grants to states for wastewater treatment plant construction under a program administered by the Environmental

Protection Agency (EPA). Federal funds are provided through annual appropriations under a state-by-state allocation formula contained in the act itself. States used their allotments to make grants to cities to build or upgrade wastewater treatment plants and thus to achieve the overall objectives of the act: restoring and maintaining the chemical, physical, and biological integrity of the nation's waters. The federal share of project costs, originally 75% under P.L. 92-500, was reduced to 55% in 1981.

By the mid-1980s, there was considerable policy debate between Congress and the Administration over the future of the act's construction grants program and, in particular, the appropriate federal role. Through FY1984, Congress had appropriated nearly $41 billion under this program, representing the largest nonmilitary public works programs since the Interstate Highway System. The grants program was a target of the Reagan Administration's budget cutters, who sought to redirect budgetary priorities in part to sort out the appropriate roles of federal, state, and local governments in a number of domestic policy areas, including water pollution control. The Administration's rationale included several points.

- The original intent of the program to address the backlog of sewage treatment needs had been virtually eliminated by the mid-1980s.
- Most remaining projects (such as small, rural systems) were believed to pose little environmental threat and were not appropriate federal responsibilities.
- State and local governments, in the Administration's view, were fully capable of running construction programs and have a clear responsibility to construct treatment capacity to meet environmental objectives that were primarily established by states.

Thus, the Reagan Administration sought a phaseout of the act's construction grants program by 1990. Many states and localities supported the idea of phasing out the grants program, since many were critical of what they viewed as burdensome rules and regulations that accompanied the federal grant money. However, they sought a longer transition and ample flexibility to set up long-term financing to promote state and local self-sufficiency.

Congress' response to this debate was contained in 1987 amendments to the act (P.L. 100-4, the Water Quality Act of 1987). It authorized $18 billion over nine years for sewage treatment plant construction, through a combination of the Title II grants program and a new State Water Pollution Control Revolving Funds (SRF) program. Under the new program, in Title VI of the act, federal grants would be provided as seed money for state-administered loans to build sewage treatment plants and, eventually, other water quality projects. Cities, in turn, would repay loans to the state, enabling a phaseout of federal involvement while the state builds up a source of capital for future investments. Under the amendments, the SRF program was phased in beginning in FY1989 (in FY1989 and FY1990, appropriations were split equally between Title II and Title VI grants) and entirely replaced the previous Title II program in FY1991. The intention was that states would have flexibility to set priorities and administer funding, while federal aid would end after FY1994.

The authorizations provided in the 1987 amendments expired in FY1994, but pressure to extend federal funding has continued, in part because, although Congress has appropriated over $77 billion in assistance since 1972, funding needs remain high: an additional $181 billion nationwide is needed for all types of projects eligible for funding under the act, according to the most recent formal estimate by EPA and states.[1] Thus, Congress has continued to appropriate funds, and the anticipated shift to full state funding responsibility has been delayed.

In contrast to the 40-plus years of federal support for financing municipal wastewater treatment facilities, Congress established a program under the Safe Drinking Water Act (SDWA) to help communities with financing of projects needed to comply with federal drinking water regulations more recently, in 1996.[2] Funding support for drinking water only occurred more recently for several reasons. First, until the 1980s, the number of drinking water regulations was fairly small, and public water systems often did not need to make large investments in treatment technologies to meet those regulations. Second and relatedly, good quality drinking water traditionally has been available to many communities at relatively low cost. By comparison, essentially all communities have had to construct or upgrade sewage treatment facilities to meet the requirements of the CWA.

Over time, drinking water circumstances have changed, as communities have grown, and commercial, industrial, agricultural, and residential land-uses have become more concentrated, thus resulting in more contaminants reaching drinking water sources. Moreover, as the number of federal drinking water standards has increased, many communities have found that their water may not be as good as once thought and that additional treatment technologies are required to meet the new standards and protect public health. Between 1986 and 1996, for example, the number of regulated drinking water contaminants grew from 23 to 83, and EPA and the states expressed concern that many of the nation's 52,000 small community water systems were likely to lack the financial capacity to meet the rising costs of complying with the Safe Drinking Water Act.

Congress responded to these concerns by enacting the 1996 SDWA Amendments (P.L. 104-182) which authorized a drinking water state revolving loan fund (DWSRF) program to help systems finance projects needed to comply with SDWA regulations and to protect public health. This program, fashioned after the Clean Water Act SRF, authorizes EPA to make grants to states to capitalize DWSRFs which states then use to make loans to public water systems. Appropriations for the program were authorized at $599 million for FY1994 and $1 billion annually for FY1995 through FY2003. According to the most recent EPAstate survey, future funding needs for projects to treat and deliver public drinking water supplies in the United States are $277 billion over the next 20 years.[3]

EPA analyses have focused attention on the issue of local governments' need for funds and what the federal role should be in assisting states and cities. In September 2002, EPA released a study, called the Gap Analysis, which assesses the difference between current capital spending for wastewater and drinking water infrastructure and total funding needs. EPA estimates that, over the next two decades, the United States needs to spend nearly $660 billion to replace existing wastewater and drinking water infrastructure systems and to build new ones. According to the study, if there is no increase in investment, there will be about a $11 billion per year gap between current capital expenditures for water infrastructure ($23 billion annually) and projected spending needs. Table 1 summarizes funding for water infrastructure programs since enactment of P.L. 100-4.

In appropriations legislation, funding for these EPA programs is contained in the measure providing funds for the Department of the Interior, Environment, and Related Agencies.[4] Within the portion of the bill which funds EPA, wastewater treatment assistance first was specified in an account called Construction Grants, which was later renamed State Revolving Funds/Construction Grants, then renamed Water Infrastructure. Since FY1996, this account has been titled State and Tribal Assistance Grants (STAG). The remainder of this report summarizes, in chronological order, congressional activity to fund items in the STAG account since the 1987 Clean Water Act amendments.

Table 1. Water Infrastructure Funding ($ in millions)

Fiscal Year	CWA Authorization	SDWA Authorization	President's Request	Appropriation
1986	$2,400		$2,400	$1,800
1987	2,400		2,000	2,361
1988	2,400		2,000	2,304
1989	2,400		1,500	1,950
1990	2,400		1,200	1,980
1991	2,400		1,600	2,100
1992	1,800		1,883	2,383
1993	1,200		2,467	2,483
1994	600	599	2,047	2,375
1995	—	1,000	2,528	2,769
1996	—	1,000	2,365	2,155
1997	—	1,000	2,178	2,201
1998	—	1,000	2,078	2,468
1999	—	1,000	2,028	2,527
2000	—	1,000	1,753	2,561
2001	—	1,000	1,753	2,621
2002	—	1,000	2,233	2,660
2003	—	1,000	2,185	2,599
2004	—	—	1,798	2,612
2005	—	—	1,794	2,336
2006	—	—	1,649	2,005
2007	—	—	1,570	2,005
2008	—	—	1,553	1,695
Total	18,000	9,599	44,562	52,950

The STAG account now includes all water infrastructure funds and management grants provided to assist states in implementing air quality, water quality, and other media-specific environmental programs. The FY1996 appropriation was the first to include both water infrastructure and other state environmental grants; the latter previously were included in EPA's general program management account. Amounts shown in **Table 1** include funds for Clean Water Act infrastructure grants, drinking water SRF grants (discussed below), and earmarked infrastructure projects grants (also discussed below), but **Table 1** does not include the funds for consolidated state environmental management grants. However, these state grants are discussed below in sections providing the appropriations chronology.

Trends in Water Infrastructure Funding

Three changes are especially evident in the recent history of water infrastructure funding, as reflected in the appropriations account where these funds are detailed. One is inclusion in the account of non-infrastructure grants to states. This began in FY1993 with addition of Clean Water Act Section 319 grants for state programs to manage nonpoint source pollution and was expanded in FY1996 to include all state grants for management of environmental programs, in a single consolidated grants appropriation.

A second trend, discussed below, has been an increasing number and amount of specially earmarked grants for needy cities and other special purpose projects. A third trend, also discussed below, is expansion of the account to include SRF capitalization grants for drinking water projects, under authority of the Safe Drinking Water Act Amendments of 1996 (P.L. 104-134).

SRF Grants vs. Special Purpose Project Grants

The practice of earmarking a portion of the construction grants/SRF account for specific wastewater treatment and other water quality projects began in the FY1989 legislation. Since then it has increased to the point of representing a significant portion of appropriated funds (31% of the total water infrastructure appropriation in FY1994, for example, 28% in FY1995, but somewhat less in recent years: 16% in FY2005 and FY2006). The number of projects receiving these earmarked funds also has increased: from four in FY1989, to eight in FY1994, 669 in FY2005, but fewer in FY2006 (259 total). Since FY2000, the larger total number of earmarked projects has resulted in more communities receiving such grants, but at the same time receiving smaller amounts of funds. Thus, while a few communities have received individual earmarked awards of $3 million or more in recent years, the average size of earmarked grants has shrunk: $18.1 million in FY1995, $4.9 million in FY1999, $600,427 in FY2005, and $1.08 million in FY2006. (Conference reports on the individual appropriations bills, noted in the later discussion in this report, provide limited detail on projects funded in this manner.) The effective result of earmarking has been to reduce the amount of funds provided to states to capitalize their SRF programs. Of the $51.3 billion appropriated to EPA for water infrastructure programs since 1986, $6.87 billion has gone to earmarked project grants.[5]

Interest groups representing state water quality program managers and administrators of infrastructure financing programs have criticized this practice of appropriators. They contend that earmarking undermines the intended purpose of the state funds, to promote water quality improvements nationwide. Many state officials would prefer that funds be allocated more equitably, not based on what they view largely as political considerations, and they would prefer that state environmental and financing officials retain responsibility to set actual spending priorities. Further, they say, because directed funding of special projects diminishes the level of seed funding to SRFs, it delays the time when states will be financially self-sufficient and may actually prolong the time when states seek continued federal support.

The practice of earmarking has been criticized because designated projects are receiving more favorable treatment than other communities' projects: they generally are eligible for 55% federal grants (and will not be required to repay 100% of the funded project cost, as is the case with a loan through an SRF), and the practice sidesteps the standard CWA process of states' determining the priority by which projects will receive funding. It also means that the projects have generally not been reviewed by the CWA authorizing committees. This is especially true since FY1992 when special purpose grant funding has been designated for projects not authorized in the Clean Water Act or amendments to it or in the Safe Drinking Water Act.

Members of Congress may intervene for a specific community for a number of reasons. In some cases, the communities may have been unsuccessful in seeking state approval to fund the project under an SRF loan or other program. For some, the cost of a project financed through a state loan, which the community must fully repay, is deemed unacceptably high, because repaying the loan can result in such increased user fees that ratepayers feel are unduly burdensome. The community then seeks a grant to avoid this costly financial scenario. This is often the case with wastewater projects in small and rural communities. A number of the special purpose grants have been made to projects characterized as "needy cities" based on local economic conditions.

In the early years of this congressional practice, special purpose grant funding originated in the House version of the EPA appropriations bill, while the Senate for the most part resisted earmarking by rejecting or reducing amounts and projects included in House-passed legislation. With this difference in legislative approach, special purpose grant funding on several occasions were an issue during the House-Senate conference on the appropriations bill. Since FY1999, however, both the House and Senate have proposed earmarked projects in their respective versions of the EPA appropriations bill, with the final total number of projects and dollar amounts being determined by conferees. In addition, as it has now been 19 years since the last major amendments to the Clean Water Act, the desire by some Members to address special needs problems that might be debated during reauthorization has increased, thus leading to greater pressure on House and Senate Members to use the appropriations process to handle such concerns.[6]

Technically, the Clean Water Act Title II grants program ended when authorizations for it expired after FY1990. One result of earmarking special purpose grants in appropriations bills has been to perpetuate grants as a method of funding wastewater treatment construction long after FY1990. At the same time, it also has resulted in Congress providing EPA grants for drinking water system projects, which had not previously been available.

Local Cost Share on Special Purpose Grants

The federal percentage share and local match required on special purpose grants depends on the project and the year of funding. For example, in the case of appropriations for projects in Boston, San Diego, New York City, and Des Moines, IA (discussed below in the section concerning FY1989), authorization of appropriations and federal cost share were specified in the 1987 Clean Water Act amendments. For a number of other projects for which appropriations were provided in FY1992 and FY1993, the appropriations acts specified that funds were provided "as grants under title II," resulting in a requirement for local communities to provide a 45% share of project costs.

After FY1993, the appropriations acts themselves are the authority for the special purpose projects grants. In the FY1995 appropriation bill, which also directed allocation of funds appropriated in FY1994 to several needy cities, Congress addressed the issue of federal and local cost shares in report language accompanying the bill, but not in the appropriation act itself.[7]

> The conferees are in agreement that the agency should work with the grant recipients on appropriate cost-share arrangements. It is the conferees' expectation that the agency will apply the 45% local cost share requirement under Title II of the Clean Water Act in most cases.

In the FY1996 appropriations, both the act and accompanying reports were silent on federal/local cost share and applicability of Title II requirements. Because of that, EPA officials planned to require only a 5% local match for most of the special purpose grants in that bill, which is the standard matching requirement for other EPA non-infrastructure grants. Under the agency's rules, the local match could include in-kind services, as well as funding toward the project.

In the FY1997 appropriations, Congress included report language as it had in FY1995 concerning federal and local cost share requirements.[8]

> The conferees are in agreement that the Agency should work with the grant recipients on appropriate cost-share agreements and to that end the conferees direct the Agency to develop a standard cost-share consistent with fiscal year 1995.

The FY1998 and FY1999 appropriations included neither bill nor report language on this point. However, language in the House and Senate Appropriations Committees' reports on the FY1998 and FY1999 bills directed EPA to work with grant recipients on appropriate cost-share arrangements.[9]

For FY2000, Congress included explicit report language concerning the local match.[10]

> The conferees agree that the $331,650,000 provided to communities or other entities for construction of water and wastewater treatment facilities and for groundwater protection infrastructure shall be accompanied by a cost-share requirement whereby 45 percent of a project's cost is to be the responsibility of the community or entity consistent with long-standing guidelines for the Agency. These guidelines also offer flexibility in the application of the cost-share requirement for those few circumstances when meeting the 45 percent requirement is not possible.

Similar report language concerning local cost-share requirements accompanied the conference reports on the appropriations bills from FY2001 through FY2005. Since FY2004, Congress has included legislative language in the appropriations act specifying the local requirement to provide 45% of a project's cost.

Since the FY2003 appropriations legislation, Congress also has specified that, except for those limited instances in which an applicant meets the criteria for a waiver of the cost-share requirement, the earmarked grant shall provide no more than 55% of an individual project's cost, regardless of the amount appropriated.

Grants for a Drinking Water SRF

One additional aspect of earmarking a portion of the account's appropriation began in FY1994 when the Administration requested and Congress agreed to provide funds to capitalize state *drinking water* SRFs. In response, Congress appropriated drinking water SRF grants three times, but those actions were contingent on enactment of authorizing legislation, which occurred in 1996, as described above. Subsequently, capitalization grants for drinking water SRF programs were provided for the first time in FY1997 and have been included in each subsequent appropriations bill. These appropriations have totaled $10.3 billion and have averaged $858 million per year. As Congress continues to support both clean water SRFs and drinking water SRFs, the two types of grants share whatever overall level of funding is available within the account.

APPROPRIATIONS CHRONOLOGY

FY1986, FY1987

The authorization period covered by P.L. 100-4 was FY1986-FY1994. By the time the amendments were enacted, FY1986 was over, as was a portion of FY1987. Thus, appropriations for those two years only indirectly reflected the policy and program changes for later years that were contained in P.L. 100-4. For FY1986, Congress appropriated a total of $1.8 billion, consisting of $600 million approved in December 1985 (while Congress was beginning to debate reauthorization legislation that eventually was enacted as P.L. 100-4 in January 1987) and $1.2 billion more in July 1986.

For FY1987, while debate on CWA reauthorization continued, President Reagan requested $2.0 billion, consistent with his legislative proposal to terminate the grants program by FY1990. In October 1986, Congress appropriated $2.4 billion (P.L. 99-500 / P.L. 99-591). However, only $1.2 billion of that amount was released immediately, pending enactment of a reauthorization bill, which was then in conference. Following enactment of the

Water Quality Act of 1987, remaining FY1987 funds were released as part of a supplemental appropriations bill (P.L. 100-71). Conferees on that measure agreed, however, to shift $39 million of the remaining unreleased grant funds to other priority water quality activities authorized in P.L. 100-4. The final total of construction grant monies was $2.361 billion.

FY1988

For FY1988 the President again requested $2.0 billion. In December 1987, Congress approved legislation providing FY1988 appropriations (P.L. 100-202, the omnibus continuing resolution to fund EPA and other federal agencies). In it, Congress appropriated $2.304 billion for construction grants. Final action on the EPA budget and other funding bills had been delayed by budget-cutting talks between Congress and the White House. Reduced construction grants funding was one of many spending cuts required to implement a congressional-White House "summit agreement" on the budget. The final construction grants appropriation was less than funding levels that had been provided in separate versions of a bill passed by the House and Senate before the budget summit, $2.4 billion.

FY1989

For FY1989, President Reagan requested $1.5 billion, or 35% below FY1988 appropriations and 37.5% less than the authorized level of $2.4 billion for FY1989. In separate versions of an EPA appropriations bill, the House and Senate voted to provide $1.95 billion and $2.1 billion respectively. The final figure, in P.L. 100-404, was $1.95 billion which included $68 million for special projects in four states. Thus, the actual amount provided for grants was $1.882 billion. That total was divided equally between the previous Title II grants program and new Title VI SRF program, as provided in the authorizing language of P.L. 100-4.

The FY1989 legislation was the first to include earmarking of funds for specified projects or grants in EPA's construction grants account, an action that continued in subsequent years, as discussed above. All of the projects funded in the 1989 legislation were ones that had been authorized in provisions of the Water Quality Act of 1987 (WQA, P.L. 100-4). The designated projects were in Boston (authorized in Section 513 of the WQA, to fund the Boston Harbor wastewater treatment project), San Diego/Tijuana (Section 510, to fund an international sewage treatment project needed because of the flow of raw sewage from Tijuana, Mexico across the border), Des Moines, IA (Section 515, for sewage treatment plant construction), and Oakwood Beach/Redhook, NY (Section 512 of the WQA, to relocate natural gas distribution facilities near wastewater treatment works in New York City).

FY1990

For FY1990, President Reagan's budget requested $1.2 billion in wastewater treatment assistance, or 50% less than the authorized level and 38.5% less than the FY1989 enacted amount of $1.95 billion. Further, the Reagan budget proposed that the $1.2 billion consist of $800 million in Title VI monies and $400 million in Title II grants, contrary to provisions of the CWA directing that appropriations be equally divided between the two grant programs, as in FY1989. President's Bush's revised FY1990 budget, presented in March 1989, made no changes from the Reagan budget in this area.

In acting on this request, Congress agreed to provide $2.05 billion, including $46 million for three special projects (Boston, San Diego/Tijuana, and Des Moines), leaving a total of $1.002 billion each for Titles II and VI (P.L. 101-144). Title II funds were reduced by $6.8 million, however, due to funds earmarked for a specific project in South Carolina. Although these amounts were appropriated, all funds in the bill were reduced by 1.55% (or, a $31.8 million reduction from the construction grants account) to provide funds for the federal government's anti-drug program.

Final FY1990 appropriations were altered again by passage of the FY1990 Budget Reconciliation measure and implementation of the Balanced Budget and Emergency Deficit Control Act (the Gramm-Rudman-Hollings Act), which established procedures to reduce budget deficits annually, resulting in a zero deficit by 1993. For each fiscal year that the deficit was estimated to exceed maximum targets established in law, an automatic spending reduction procedure was triggered to eliminate deficits in excess of the targets through "sequestration," or permanent cancellation of budgetary resources.

Thus, to meet budget reduction mandates and, in particular, deficit reduction targets under the Balanced Budget and Emergency Deficit Control Act (the Gramm-Rudman-Hollings Act), additional funding cuts were included in P.L. 101-239, the Budget Reconciliation Act of 1989, affecting construction grants funding and all other accounts not exempted from Gramm-

Rudman procedures. P.L. 101-239 provided that the "sequestration" procedures under the Gramm-Rudman-Hollings Act would be allowed to apply for a portion of FY1990 (for 130 days, or 35.6% of the year), providing an additional automatic spending reduction in EPA and other agencies' programs subject to the act.

As a result of these reductions, funding for wastewater treatment aid in FY1990 totaled $1.98 billion, or $30 million more than in FY1989. The total included $53 million for special projects in San Diego, Boston, Des Moines, and Honea Path/Ware Shoals, SC, $960 million for Title II grants, and $967 million for Title VI grants. The combined reductions amounted to 3.4% less than the amount agreed to by conferees on P.L. 101-144 (i.e., $2.05 billion), before subtracting funds for anti-drug programs and accounting for effects of the Gramm-Rudman partial-year sequester.

FY1991

For FY1991, President Bush requested $1.6 billion in funding for wastewater treatment assistance. This total included $15.4 million to be earmarked for the San Diego project authorized in Section 510 of the Water Quality Act of 1987, to fund construction of an international sewage treatment project. The remainder, $1.584 billion, would be only for capitalization grants under Title VI of the act, as the 1987 legislation provides for no new Title II grants after FY1990.

In acting on EPA's appropriations for FY1991 (P.L. 101-507), Congress agreed to provide $2.1 billion in wastewater treatment assistance. Beginning in FY1991, all appropriated funds are utilized for capitalization grants under Title VI of the act (as provided in the Water Quality Act of 1987); funding for the traditional Title II grants program was no longer available.

The enacted level included several earmarkings: $15.7 million for San Diego, $20 million for Boston Harbor (section 513 of the WQA), and $16.5 million for a new Water Quality Cooperative Agreement Program under Section 104(b)(3) of the act.[11] The President's budget had requested $16.5 million to support state permitting, enforcement and water quality management activities, especially to offset the reductions in aid to states due to elimination of state management setasides from the previous Title II construction grants program. Congress agreed to the level requested, but provided it as a portion of the wastewater treatment appropriation, rather than as part of EPA's general program management appropriation, as in the

President's request. As a result of these earmarkings, $2.048 billion was provided for Title VI grants.

FY1992

For FY1992, President Bush requested $1.9 billion in wastewater treatment funds, or $100 million more than authorized under the Water Quality Act of 1987 for Title VI grants in FY1992. However, out of the $1.9 billion total, the President's request sought $1.5 billion for Title VI SRF grants and $400 million as grants under the expired Title II construction grants program for the following coastal cities: Boston, San Diego, New York, Los Angeles, and Seattle. Two of the five designated projects had been authorized in the 1987 Clean Water Act amendments; the other three did not have explicit statutory authorization. Also, $16.5 million was requested for Water Quality Cooperative Agreement grants to the states.

In acting on the request in November 1991, Congress provided total wastewater funds of $2.4 billion (P.L. 102-139). The total was allocated as follows:

- $1,948.5 million for SRF capitalization grants,
- $16.5 million for section 104(b)(3) grants,
- $49 million for the special project in San Diego-Tijuana (section 510 of the Water Quality Act),
- $46 million to the Rouge River (MI) National Wet Weather Demonstration Project, and
- $340 million as construction grants under title II of the Clean Water Act for several other special projects — the Back River Wastewater Treatment Plant (Baltimore), Maryland, the Boston Harbor project, New York City, Los Angeles, San Diego (a wastewater reclamation project), and Seattle.

This appropriation bill was the first to include special purpose grant funding for several projects not specifically authorized in the Clean Water Act or amendments to that law.

FY1993

For FY1993, President Bush requested $2.484 billion for state revolving funds/construction grants (now called the water infrastructure account). The requested total included $340 million to be targeted for 55% construction grants to six communities: Boston, New York, Los Angeles, San Diego, Seattle, and Baltimore. In addition, the President requested that $130 million be directed toward a Mexican Border Initiative, consisting of $65 million for construction of the international treatment plant at San Diego (to address the Tijuana sewage problem), $15 million for projects at Nogales AZ and New River, CA, and $50 million as 50% grants for *colonias* in Texas.[12] The President also requested $16.5 million for Section 104(b)(3) grants. Along with these special project and grant amounts, the request sought $2.014 billion for SRF assistance.

Final action on FY1993 funding occurred on September 25, 1992 (P.L. 102-389). It provided an appropriation of $2.55 billion, but $622.5 million of this amount was reserved for special projects and other grants. The bill provided $50 million in CWA Section 319 grants[13] and $16.5 million in Section 104(b)(3) grants out of the SRF amount. It included $556 million for the following special purpose grants: the international treatment plant at San Diego (Tijuana — Section 510 of the WQA, with bill language capping funding for that project at $239.4 million), plus projects in Boston, New York, Los Angeles, San Diego, Seattle, Rouge River MI, Baltimore, Ocean County NJ, Atlanta, and for *colonias* in Texas, Arizona and New Mexico. The final SRF grant amount under the bill was $1.928 billion.

Early in 1993, President Clinton requested that Congress approve "economic stimulus and investment" spending, in the form of supplemental FY1993 appropriations. Both his original proposal and a subsequent modified proposal included additional SRF grant funds, but neither of the bills enacted by Congress in response to these requests (P.L. 103-24, P.L. 103-50) provided additional SRF funds.

FY1994

For FY1994, the Clinton Administration requested $2.047 billion for water infrastructure. The funds in this request were $1.198 billion to capitalize State Revolving Funds, $150 million for Mexican Border Project grants, and

$100 million for a single hardship community (Boston). The request also included $599 million to capitalize new state drinking water revolving funds.

The final version of the FY1994 legislation (P.L. 103-124) provided $2.477 billion for water infrastructure/state revolving funds. Of this total amount, $599 million was to be reserved for drinking water SRFs, if authorization legislation were enacted; $80 million was for Section 319 grants; $22 million was for Section 104(b)(3) grants; and $58 million was for Tijuana/San Diego-Section 510 of the WQA. This resulted in an appropriation of $1.718 billion for clean water SRFs.

In addition, the final bill provided that $500 million be used to support water infrastructure financing in economically distressed/hardship communities. Under the bill, these funds were not available for spending until May 31, 1994, and were set aside until projects were authorized in the CWA for this purpose.

Thus, the bill as enacted provided $1.218 billion immediately for clean water SRFs, with the expectation that $500 million more would be available for financing hardship community projects after May 31, 1994.

FY1995

For FY1995, President Clinton requested $2.65 billion for water infrastructure consisting of: $1.6 billion for CWA SRFs, $100 million for Section 319 nonpoint source management grants to states, $52.5 million for a grant to San Diego for a wastewater project pursuant to Section 510 of the WQA, $47.5 million for other Mexican border projects, $50 million to the State of Texas for *colonias* projects, and $100 million for grants under Title II for needy cities (intended for Boston). The request included $700 million for drinking water SRFs, pending enactment of authorizing legislation. The President's budget also requested $21.5 million for Section 104(b)(3) grants/cooperative agreements.

Final agreement on FY1995 funding was contained in P.L. 103-327, enacted in September 1994, which provided a total of $2.962 billion for water infrastructure financing. Of the total, $22.5 million was for grants under Section 104(b), $100 million for Section 319 grants, $70 million for public water system grants (grants to states under the Safe Drinking Water Act to support state implementation of delegated drinking water programs), $52.5 million for the Section 510 project in San Diego, and $700 million for drinking water SRFs (contingent upon enactment of authorization legislation).

The remaining $2.017 billion was for CWA projects. Of this amount, $1.235 billion was for clean water SRF grants to states under Title VI of the CWA. The remaining $781.8 million (39% of this amount, 26% of the total appropriation) was designated for 45 specific, named projects in 22 states. The earmarked amounts ranged in size from $200,000 for Southern Fulton County, PA, to $100 million for the City of Boston.

Finally, the conferees included bill language concerning release of the $500 million in FY1994 needy cities money (because the authorizing committees of Congress had not acted on legislation to authorize specific projects, as had been intended in P.L. 103-124) as follows:

- $150 million to Boston, $50 million for *colonias* in Texas, $10 million for *colonias* in New Mexico, $70 million for a New York City wastewater reclamation facility, $85 million for the Rouge River project, $50 million for the City of Los Angeles, $50 million for the County of Los Angeles, and $35 million for Seattle, WA.

FY1996

In February 1995, President Clinton submitted the Administration's budget request for FY1996. It requested $2.365 billion for water infrastructure funding consisting of $1.6 billion for clean water state revolving funds, $500 million for drinking water state revolving funds, $150 million to support Mexico border projects under the U.S.-Mexican Border Environmental Initiative and NAFTA and $100 million for special need/economically distressed communities (not specified in the request, but presumed to be intended for Boston), plus $15 million for water infrastructure needs in Alaska Native Villages.

In February 1995, congressional appropriations committees began considering legislation to rescind previously appropriated FY1995 funds, as part of overall efforts by the 104th Congress to shape the budget and federal spending. These efforts resulted in passage in July of P.L. 104-19 which rescinded $16.5 billion in total funds from a number of departments, agencies, and programs. In the water infrastructure area, it rescinded $1,077,200,000 from prior year appropriations including the $3.2 million for a project in New Jersey (it had mistakenly been funded twice in P.L. 103-327) and $1,074,000,000. Although not contained in bill language, it was understood that the larger rescinded amount consisted solely of drinking water SRF funds (leaving $1.235 billion for FY1995 clean water SRF funds, $778.6 million for earmarked wastewater projects — both amounts as originally appropriated — and $225 million in FY1994-FY1995 drinking water SRF funds that had not yet been authorized).

It took until April 1996 for Congress and the Administration to reach agreement on FY1996 appropriations for EPA as part of omnibus legislation (P.L. 104-134) that consolidated five appropriations bills not yet enacted due to disagreements over funding levels and policy. Agreement came as the fiscal year was more than one-half over.

Before that, however, congressional conferees reached agreement in November 1995 on FY1996 legislation for EPA (H.R. 2099, H.Rept. 104-353). Conferees agreed to provide $2.323 billion for a new account titled State and Tribal Assistance Grants, consisting of infrastructure assistance and state environmental management grants for 16 categorical programs that had previously been funded in a separate appropriations account. The total included $1.125 billion for clean water SRF grants, $275 million in new appropriations for drinking water SRF grants, and $265 million for special purpose project grants. Report language provided that the drinking water SRF money also included $225 million from FY1995 appropriations that were rescinded in P.L. 104-19. The drinking water SRF money would be available upon reauthorization of the Safe Drinking Water Act; otherwise, it would revert to clean water SRF grants if the SDWA were not reauthorized by June 30, 1996. This made the total for drinking water SRF grants $500 million.

The November 1995 agreement on H.R. 2099 included $658 million for consolidated state environmental grants. In doing so, Congress endorsed an Administration proposal for a more flexible approach to state grants, a key element of EPA's efforts to improve the federal-state partnership in environmental programs. In lieu of traditional grants provided separately to support state air, water, hazardous waste and other programs, consolidated grants are intended to reduce administrative burdens and improve environmental performance by allowing states and tribes to target funds to meet their specific needs and integrate their environmental programs, as appropriate. Congress' support was described in accompanying report language.[14]

> The conferees agree that Performance Partnership Grants are an important step to reducing the burden and increasing the flexibility that state and tribal governments need to manage and implement their environmental protection programs. This is an opportunity to use limited resources in the

> most effective manner, yet at the same time, produce the results-oriented environmental performance necessary to address the most pressing concerns while still achieving a clean environment.

Including state environmental grants in the same account with water infrastructure assistance reflects Congress' support for enhancing the ability of states and localities to implement environmental programs flexibly and support for EPA's ability to provide block grants to states and Indian tribes.

The H.R. 2099 conference agreement also included legislative riders intended to limit or prohibit EPA from spending money to implement several environmental programs. The Administration opposed the riders. The House and Senate approved this bill in December, but President Clinton vetoed it, because of objections to spending and policy aspects of the legislation.

With no full-year funding in place from October 1995 to April 1996, EPA and the programs it administers (along with agencies and departments covered by four other appropriations bills not yet enacted) were subject to a series of short-term continuing resolutions, some lasting only a day, some lasting several weeks. In March 1996, the House and Senate began consideration of an omnibus appropriations bill to fund EPA and other agencies for the remainder of FY1996, finally reaching agreement in April on a bill (H.R. 3019) enacted as P.L. 104-134.[15] Congress agreed to provide $2.813 billion for a new account titled State and Tribal Assistance Grants, consisting of state grants and infrastructure assistance, as in H.R. 2099, the vetoed measure. The total was divided as follows:

- $658 million for consolidated state environmental grants,
- $1.3485 billion for clean water SRF grants (including $50 million for impoverished communities),
- $500 million in new appropriations for drinking water SRF grants,
- $150 million for Mexico-border project grants and Texas *colonias*, as requested,
- $15 million for Alaskan Native Villages, as requested, and
- $141.5 million for 17 special purpose project grants.

Report language provided that the drinking water SRF money also included $225 million from FY1995 appropriations that remained available after the rescissions in P.L. 104-19, for a total of $725 million. The drinking water SRF money was contingent upon reauthorization of the Safe Drinking

Water Act by August 1, 1996; otherwise, it would revert to clean water SRF grants.

The final agreement (P.L. 104-134) included several of the legislative riders from previous versions of the legislation, including riders related to drinking water and clean air, but dropped others strongly opposed by the Administration.

Funds within the State and Tribal Assistance Grants account were redistributed after Congress passed Safe Drinking Water Act amendments in August 1996. Enactment of the amendments (P.L. 104-182) occurred on August 6 — after the August 1 deadline in P.L.104-134 that would have made $725 million available for drinking water SRF grants in FY1996. Thus, the previously appropriated $725 million reverted to clean water SRF grants, making the FY1996 total for those grants $2.0735 billion.

FY1997

While debate over the FY1996 appropriations was continuing, in March 1996, President Clinton submitted the details of a FY1997 budget. For water infrastructure and state and tribal assistance, the request totaled $2.852 billion consisting of

- $1.35 billion for clean water SRF grants (the request included language that would authorize states the discretion to use this SRF money either for clean water or drinking water projects),
- $165 million for US-Mexico border projects, Texas *colonias*, and Alaskan Native Village projects,
- $113 million for needy cities projects,
- $674 million for state performance partnership consolidated management grants, and
- $550 million for drinking water infrastructure SRF funding, contingent upon enactment of authorizing legislation.

In response to the Administration's request, in June 1996 the House approved legislation (H.R. 3666) providing FY1997 funding for EPA. In the State and Tribal Assistance Grants account, the House approved $2.768 billion, $84 million less than requested but on the whole endorsing the budget request. The total provided the following: $1.35 billion for clean water SRF grants, as requested; $165 million, as requested, for U.S.-Mexico border projects, Texas *colonias*, and Alaskan Native Village projects; $450 million for drinking water SRF funding, contingent upon authorization; $674 million for state performance partnership consolidated management grants; and $129 million for seven special purpose grants.

In July, the Senate Appropriations Committee reported its version of H.R. 3666. The committee approved $2.815 billion for this account, consisting of $1.426 billion for clean water SRF grants; $550 million for drinking water SRF grants, contingent upon authorization; $165 million, as requested, for U.S.-Mexico border projects, Texas *colonias*, and Alaskan Native Village projects; and $674 million for consolidated state grants. The committee rejected the provision of the House-passed bill providing $129 million for special purpose grants, including funds for Boston and New Orleans requested by the Administration, saying in report language that earmarking is provided at the expense of state revolving funds and does not represent an equitable distribution of grant funds (S.Rept. 104-318).

During debate on H.R. 3666 in September, the Senate adopted an amendment to reduce the FY1997 appropriation for clean water SRF grants by $725 million in order to fund the new drinking water SRF program. This action was intended to restore funds to the drinking water program which had been lost when Safe Drinking Water Act amendments were not enacted by August 1, 1996. Thus, the Senatepassed bill provided $701 million for clean water SRF grants and $1.275 billion for drinking water SRF grants for FY1997. Other amounts in the account were unchanged.

The conference report on H.R. 3666 (H.Rept. 104-812) was approved by the House and Senate on September 24, 1996. President Clinton signed the bill September 26 (P.L. 104-204). It reflected compromise of the House- and Senate-passed bills, providing the following amounts within the State and Tribal Assistance Grants account ($2.875 billion total):

- $625 million for clean water SRF grants,
- $1.275 billion for drinking water SRF grants,
- $165 million, as requested, for U.S.-Mexico border projects, Texas *colonias*, and Alaskan Native Village projects,
- $674 million for consolidated state grants, and
- $136 million for 18 specific wastewater, water, and groundwater project grants (the 7 specified in House-passed H.R. 3666, plus 11 more; the bill provided funds for each of the needy cities projects requested by the Administration, but in lesser amounts).

The allocation of clean water and drinking water SRF grants was consistent with the Senate's action to restore funds to the drinking water program after enactment of the Safe Drinking Water Act amendments in early August.

Subsequently, Congress passed a FY1997 Omnibus Consolidated Appropriations bill to cover agencies and departments for which full-year funding had not been enacted by October 1, 1996 (P.L. 104-208). It included additional funding for several EPA programs, as well as $35 million (on top of $40 million provided in P.L. 104-204) for the Boston Harbor cleanup project.

FY1998

President Clinton presented the Administration's budget request for FY1998 in February 1997. For water infrastructure and state and tribal assistance, the request totaled $2.793 billion, consisting of $1.075 billion for clean water SRF grants, $725 million for drinking water SRF grants, $715 million for consolidated state environmental grants, and $278 million for special project grants.

House and Senate committees began activities on FY1998 funding bills somewhat late in 1997, due to prolonged negotiations between Congress and the President over a five-year budget plan to achieve a balanced budget by 2002. After appropriators took up the FY1998 funding bills in June, the House passed EPA's appropriation in H.R. 2158 (H.Rept. 105-175) on July 15. In the State and Tribal Assistance Grants account, the House approved $3.019 billion, consisting of $1.25 billion for clean water SRF grants ($600 million more than FY1997 levels and $175 million more than requested by the President), $750 million for drinking water SRF grants ($425 million less than FY1997 levels, but $25 million more than the request), $750 million for state environmental assistance grants, and $269 million for special projects. The latter included funds for the special projects requested by the Administration but at reduced levels ($149 million total for these projects), plus $120 million in special project grants for 21 other communities.

The Senate passed a separate version of an FY1998 appropriations bill on July 22, 1997 (S. 1034, S.Rept. 105-53). It provided $3.047 billion for the STAG account, consisting of $1.35 billion for clean water SRF grants, $725 million for drinking water SRF grants, $725 million for state environmental assistance grants, and $247 million for special project grants. The Senate bill provided the amounts requested by the Administration for U.S.-Mexico border projects, Texas *colonias*, and Alaskan Native Village projects (but no special funds for others requested by the President), plus $82 million for 18 special project grants for other communities identified in report language.

Conferees reached agreement on FY1998 funding in early October 1997 (H.R. 2158, H.Rept. 105-297). The final version passed the House on October 8 and passed the Senate on October 9. President Clinton signed the bill October 27 (P.L. 105-65). As enacted, it provided $3.213 billion for the STAG account, consisting of $1.35 billion for clean water SRF grants, $725 million for drinking water SRF grants, $745 million for consolidated state environmental assistance grants, and $393 million for 42 special purpose project and special community need grants for construction of wastewater, water treatment and drinking water facilities, and groundwater protection infrastructure. It included the following amounts for grants requested by the Administration:

- $75 million for U.S.-Mexico border projects,
- $50 million for Texas *colonias*,
- $50 million for Boston Harbor wastewater needs,
- $10 million for New Orleans,
- $3 million for Bristol County, MA, and
- $15 million for Alaskan Native Village projects.

The final bill also provided funds for all of the special purpose projects included in the separate House and Senate versions of the legislation, plus three projects not included in either earlier version.

Bill language was included in P.L. 105-65 to allow states to cross-collateralize clean water and drinking water SRF funds, that is, to use the combined assets of amounts appropriated to State Revolving Funds as common security for both SRFs, which conferees said is intended to ensure maximum opportunity for states to leverage these funds. Senate committee report language also said that the conference report on the 1996 Safe Drinking Water Act Amendments had stated that bond pooling and similar arrangements were not precluded under that legislation. The appropriations bill language was intended to ensure that EPA does not take an unduly narrow interpretation of this point which would restrict the states' use of SRF funds.[16]

On November 1, 1997, President Clinton used his authority under the Line Item Veto Act (P.L. 104-130) to cancel six items of discretionary budget authority provided in P.L. 105-65. The President's authority under this act took effect in the 105th Congress; thus, this was the first EPA appropriations bill affected by it. The cancelled items included funding for one of the special purpose grants in the bill, $500,000 for new water and sewer lines in an industrial park in McConnellsburg, PA. Reasons for the cancellation, according to the President, were that the project had not been requested by the Administration; it would primarily benefit a private entity and is outside the scope of EPA's usual mission; it is a low priority use of environmental funds; and it would provide funding outside the normal process of allocating funds according to state environmental priorities.[17]

However, in June 1998, the Supreme Court struck down the Line Item Veto Act as unconstitutional, and in July the Office of Management and Budget announced that funding would be released for 40-plus cancellations made in 1997 under that act (including those cancelled in P.L. 105-65) that Congress had not previously overturned. (For additional information, see CRS Report RL33635, *Item Veto and Expanded Impoundment Proposals: Legislative History and Current Status*.)

FY1999

President Clinton's budget request for FY1999, presented to Congress in February 1998, requested $2.9 billion for the State and Territorial Assistance Grants account, representing 37% of the $7.9 billion total requested for EPA programs. The total included $1.075 billion for clean water SRF grants, $775 million for drinking water SRF grants, $115 million for water infrastructure projects along the U.S.- Mexico border projects and in Alaskan Native Villages, $78 million for needy cities projects, and $875 million for consolidated state environmental grants.

Legislative action on the budget request occurred in mid-1998. Both houses of Congress increased amounts for water infrastructure financing, finding the Administration's request for clean water and drinking water SRF grants, as well as special project funding, not adequate. First, the Senate Appropriations Committee reported its version of an EPA spending bill in June (S. 2168, S.Rept. 105-216). This bill, passed by the Senate July 17, provided $3.2 billion for the STAG account, consisting of $1.4 billion for clean water SRF grants, $800 million for drinking water SRF grants, $105 million for U.S.-Mexico and Alaskan Native Village projects, $100 million for 39 other special needs infrastructure grants, and $850 million for state performance partnership/categorical grants. As in FY1998, the committee included bill language allowing states to cross-collateralize their clean water and drinking water State Revolving Funds, making the language explicit for FY1999 and thereafter.

Second, the House passed its version of EPA's funding bill (H.R. 4194, H.Rept. 105-610) on July 29. This bill provided $3.2 billion for the STAG account, consisting of $1.25 billion for clean water SRF grants, $775 million for drinking water SRF grants, $70 million for U.S.-Mexico and Alaskan Native Village projects, $253.5 million for 49 other special needs infrastructure grants (including nine projects also funded in the Senate bill), and $885 million for state environmental management grants (a 20% increase above FY1998 amounts for these state grants).

Conferees resolved differences between the two versions in October 1998 (H.R. 4194, H.Rept. 105-769). The conference agreement provided $3.4 billion for the STAG account, consisting of $1.35 billion for clean water SRF grants, $775 million for drinking water SRF grants, $80 million for U.S.-Mexico and Alaskan rural and Native Village projects, $301.8 million for 80 other special needs project grants, and $880 million for state and tribal environmental program grants. The House and Senate approved the agreement on October 7 and 8, respectively, and President Clinton signed the bill into law on October 21 (P.L. 105-276).

Additional funding was provided in the Omnibus Consolidated and Supplemental Appropriations Act, FY1999 (P.L. 105-277). This bill, which provided full-year funding for agencies and departments covered by seven separate appropriations measures, directed $20 million more in special needs grants for the Boston Harbor wastewater infrastructure project, on top of $30 million that was included in P.L. 105-276.

FY2000

For FY2000, beginning on October 1, 1999, the Administration requested $2.638 billion for water infrastructure assistance and state environmental grants. The total, $370 million less than the FY1999 appropriation for this account, consisted of $800 million for clean water SRF grants, $825 million for drinking water SRF grants, $128 million for Mexican Border and special project grants, and $885 million for consolidated state environmental grants.

The request included one SRF policy issue. The Administration asked the appropriators to grant states the permission to set aside up to 20% of FY2000 clean water SRF monies in the form of grants for local communities to implement nonpoint source pollution and estuary management projects.

Currently, under the Clean Water Act, SRFs may only be used to provide loans. Some have argued that some types of water pollution projects which are eligible for SRF funding may not be suitable for loans, as they may not generate revenues which can be used to repay the loan to a state. This new authority, the Administration said, would allow states greater flexibility to address nonpoint pollution problems. Critics of the proposal said that making grants from an SRF would reduce the long-term integrity of a state's fund, since grants would not be repaid.

Members of Congress and stakeholder groups were particularly critical of the budget request for clean water SRF grants, $550 million (40%) less than the FY1999 level. Critics said the request was insufficient to meet the needs of states and localities for clean water infrastructure. In response, EPA acknowledged that several years ago the Administration made a commitment to states that the clean water SRF would revolve at $2 billion annually in the year 2005. Because of loan repayments and other factors, EPA said, the overall fund will be revolve at $2 billion per year in the year 2002, even with the 20% grant setaside included in the FY2000 request. According to EPA, the $550 million decrease from 1999 would have only a limited impact on SRFs and would still allow the Agency to meet its long-term capitalization goal of providing an average amount of $2 billion in annual assistance.

The House and Senate passed their respective versions of an EPA appropriations bill (H.R. 2684) in September 1999. The conference committee report resolving differences between the two versions (H.Rept. 106-379) was passed by the House on October 14 and the Senate on October 15 and was signed by the President on October 20 (P.L. 106-74). The final bill provided $7.6 billion overall for EPA programs, including $3.47 billion for the STAG account. Within that account, the bill included $1.35 billion for clean water SRF grants, $820 million for drinking water SRF grants, $885 million for consolidated state grants, $80 million for U.S.-Mexico Border and Alaska Rural and Native Village projects, and $331.6 million for 141 other special needs water and wastewater grants specified in report language. The final bill did not approve the Administration's request to allow states to use up to 20% of clean water SRF monies as grants for nonpoint pollution and estuary management projects.

Subsequent to enactment of the EPA funding bill, Congress passed the Consolidated Appropriations Act for FY2000 with funding for five other agencies (P.L. 106-113), which included provisions requiring a government-wide cut of 0.38% in discretionary appropriations. The bill gave the President some flexibility in applying this across-the-board reduction. Details of the reduction were announced at the time of the release of the FY2001 budget. EPA's distribution of the rescission resulted in a total reduction of $16.3 million for 139 of the special needs water and wastewater projects identified in P.L. 106-74. These projects were reduced 4.9% below enacted levels. The Agency did not reduce funds for the two projects that had been included in the President's FY2000 budget request (Bristol County, MA and New Orleans, LA) or for the United States-Mexico Border and the Alaska Rural and Native Villages programs. EPA also reduced funds for the clean water SRF (enacted at $1.35 billion) by 0.3%, for a final funding level of $1.345 billion. The drinking water SRF appropriation was not reduced, nor were consolidated state grants.

FY2001

The President's budget for FY2001 requested a total of $2.9 billion for EPA. For the second year in a row, President Clinton requested $800 million for the clean water SRF program, a $545 million reduction from the FY2000 level. The request included $825 million for the drinking water SRF program, $100 million for U.S.- Mexico Border project grants, $15 million for Alaskan Native Villages projects, two needy cities grants totaling $13 million (Bristol County, MA and New Orleans, LA), plus $1.069 billion for consolidated state environmental grants.

The budget included a policy request similar to one in the FY2000 budget, which Congress rejected. The FY2001 budget sought flexibility for states to set aside up to 19% of clean water SRF monies in the form of grants for local communities to implement nonpoint source pollution and estuary management projects.

The House approved its version of EPA's funding bill (H.R. 4635, H.Rept. 106-674) on June 21, 2000. For the STAG account, H.R. 4635 provided $3.2 billion ($273 million more than requested, but $288 million below the FY2000 level). The total in the STAG account consisted of $1.2 billion for clean water SRF grants, $825 million for drinking water SRF grants, $1.068 billion (the budget request) for categorical state grants, and $85 million for U.S.-Mexico Border and Alaska Rural and Native Villages projects. Beyond these, however, the House-passed bill included no funds for other special needs grants.

The Senate approved its version of the funding bill (S.Rept. 106-410) on October 12. For the STAG account, the Senate-passed bill provided $3.3

billion, consisting of $1.35 billion for clean water SRF grants, $820 million for drinking water SRF grants, $955 million for categorical state grants, $85 million for U.S.- Mexico Border and Alaska Rural and Native Village projects, and $110 million for special needs water and wastewater grants.

In October, the House and Senate approved EPA's funding bill for FY2001 (H.Rept. 106-988), providing $1.35 billion for clean water SRF grants (the same level enacted for FY2000) and $825 million for drinking water SRF grants. The enacted bill included $110 million for water infrastructure project grants in Alaskan Rural and Native Villages and U.S.-Mexico border projects and an additional $336 million for 237 other specified project grants throughout the country. The bill also provided $1,008 million for state categorical program grants ($60 million less in total than requested). Total funding for the STAG account was $3.6 billion. Congress disapproved the Administration's policy request concerning use of clean water SRF monies. President Clinton signed the bill October 27 (P.L. 106-377).

Subsequently, in December, Congress provided $21 million more for five more special project water infrastructure grants (in addition to the $336 million in P.L. 106-377) as a provision of H.R. 4577, the FY2001 Consolidated Appropriations bill (P.L. 106-554). Also in that legislation, Congress enacted the Wet Weather Water Quality Act, authorizing a two-year, $1.5 billion grants program to reduce wet weather flows from municipal sewer systems. The provision was included in Section 112, Division B, of P.L. 106-554.

FY2002

In April 2001, the Bush Administration presented its budget request for FY2002. The Administration requested a total of $2.1 billion for clean water infrastructure funds, consisting of $823 million for drinking water SRF grants, $850 million for clean water SRF grants (compared with $1.35 billion appropriated for FY2001), and $450 million for the new program of municipal sewer overflow grants under legislation enacted in December, the Wet Weather Water Quality Act. However, that act provides that sewer overflow grants are only available in years when at least $1.35 billion in clean water SRF grants is appropriated. Subsequently, Administration officials said they would request that Congress modify the provision linking new grant funds to at least $1.35 billion in clean water SRF grants. The Bush budget requested no funds for special earmarked grants, except for $75 million to fund projects along the U.S.-Mexico border and $35 million for projects in Alaskan Native Villages (both are the same amounts provided in FY2001). In response, some Members of Congress and outside groups criticized the budget request, saying that it did not provide enough support for water infrastructure programs. The President's budget also requested $1.06 billion for state categorical program grants.

The House passed its version of FY2002 funding for EPA on July 30 (H.R. 2620, H.Rept. 107-159). The House-passed bill provided a total of $2.4 billion for water infrastructure funds, consisting of $1.2 billion for clean water SRF grants, $850 million for drinking water SRF grants, $200 million for special project grants (individual projects were unspecified in the report accompanying H.R. 2620), $75 million for U.S.-Mexico Border projects, and $30 million for Alaskan Rural and Native Villages. The House bill provided no separate funds for the new wet weather overflow grant program, which the Administration had requested. Including $1.08 billion for state categorical program grants, total STAG account funding in the bill was $3.44 billion, about $150 million higher than the President's request.

The Senate passed its version of this appropriations bill on August 2 (S. 1216, S.Rept. 107-43). Like the House, the Senate rejected separate funding for wet weather overflow grants, and the Senate increased clean water SRF grant funding to the FY2001 level. The Senate-passed total for the STAG account was $3.49 billion, including $1.35 billion for clean water SRF grants, $850 million for drinking water SRF grants, $140 million for special needs infrastructure grants specified in accompanying report language, $75 million for U.S.-Mexico Border projects, $30 million for Alaskan Rural and Native Villages, and $1.03 billion for state categorical program grants.

Resolution of this and other appropriations bills in fall 2001 was complicated by congressional attention to general economic conditions and responses to the September 11 terrorist attacks on the World Trade Center and the Pentagon. Nevertheless, the House and Senate gave final approval to legislation providing EPA's FY2002 funding (H.R. 2620, H.Rept. 107-272) on November 8, and President Bush signed the bill on November 26 (P.L. 107-73). The final bill did not include separate funds for the new sewer overflow grant program requested by the Administration, which both the House and Senate had rejected, but it did include $1.35 billion for clean water SRF grants, $850 million for drinking water SRF grants, $344 million for 337 earmarked water infrastructure project grants specified in report language, and the requested $75 million for U.S.-Mexico Border projects and $30 million for

Alaskan Rural and Native Villages. The bill included total STAG funding of $3.7 billion.

FY2003

President Bush presented the Administration's FY2003 budget request in February 2002, asking Congress to appropriate $2.185 billion for EPA's water infrastructure programs (compared with $2.659 billion appropriated for FY2002). The FY2003 request sought $1.212 billion for clean water SRF grants, $850 million for drinking water SRF grants, and $123 million for a limited number of special projects (especially in Alaska Native Villages and in communities on the U.S.- Mexico border). The Administration proposed to eliminate funds for unrequested infrastructure project spending that Congress had earmarked in the FY2002 law, which totaled $344 million. Also, the Administration requested no funds for the municipal sewer overflow grants program enacted in 2000.

Members of Congress criticized the request level for clean water SRF capitalization grants, which was $138 million below the FY2002 enacted amount. In August 2002, the Senate Appropriations Committee approved an FY2003 funding bill for EPA that would provide $1.45 billion for clean water SRF grants, $100 million more than the FY2002 level (S. 2797, S.Rept. 107-222). In addition, the Senate committee bill included $875 million for drinking water SRF grants, $140 million for special needs infrastructure grants specified in report language, $45 million for Alaskan Rural and Native Village project grants, $75 million for U.S.- Mexico Border projects, and $1.134 billion for state categorical program grants.

The House Appropriations Committee approved its version of an FY2003 funding bill with $1.3 billion for the clean water SRF program (H.R. 5605, H.Rept. 107-740) in October. This bill also included $850 million for drinking water SRF grants, $227.6 million for special needs infrastructure grants enumerated in report language, $35 million for Alaskan Rural and Native Village project grants, $75 million for U.S.-Mexico Border projects, and $1.173 billion for state categorical program grants. Neither appropriations committee included funds for the sewer overflow grant program authorized in 2000 (the Administration did not request FY2003 funds for these grants).

Due to complex budgetary disputes during the year, final action did not occur before the 107th Congress adjourned in November 2002, and it extended into 2003, more than five months after the start of the fiscal year. Congress and the President reached agreement on funding levels for EPA and other non-defense agencies in omnibus appropriations legislation, H.J.Res. 2 (H.Rept. 108-10), which the President signed on February 20 (P.L. 108-7). The EPA portion of the enacted bill included $1.34 billion for clean water SRF grants, $844 million for drinking water SRF grants, and $413 million more for 489 special water infrastructure project grants to individual cities specified in conference report language, plus projects in Alaska Native Villages and communities on the U.S.-Mexico border. It also provided a total of $1.14 billion for categorical state grants.[18]

FY2004

On February 3, 2003, before completion of the FY2003 appropriations, President Bush submitted his budget request for FY2004. It requested a total of $1.798 billion for water infrastructure funds, consisting of $850 million for clean water SRF grants, $850 million for drinking water SRF grants, and $98 million for priority projects (especially in Alaska Native Villages and in communities on the U.S.-Mexico border). As in previous years, the Administration requested no funds for congressionally earmarked project grants for individual communities. Members of Congress and interest groups criticized the request for clean water SRF grants ($490 million below the FY2003 enacted level), but Administration officials responded by saying that the request reflects a commitment to fund this program at the $850 million level through FY2011. Funding at that level and over that longterm period, plus repayments of previous SRF loans made by states, is expected to increase the revolving levels of the overall program from $2.0 billion to $2.8 billion per year, the Administration said. The President's budget also requested $1.2 billion for categorical state grants.

On July 25, the House approved H.R. 2861 (H.Rept. 108-235), providing FY2004 appropriations for EPA. As passed, the bill included $1.2 billion for clean water SRF grants, $850 million for drinking water SRF grants, $203 million for earmarked water infrastructure project grants, and $75 million in grants for highpriority projects in Alaskan Native Villages and along the U.S.-Mexico border. Senate action on its version of a funding bill for EPA (S.Rept. 108-143) occurred on November 18. The Senate-passed bill provided $1.35 billion for clean water SRF grants, $850 million for drinking water SRF grants, $130 million for targeted infrastructure project grants, plus $95 million in grants for projects in Alaskan Native Villages and along the U.S.-Mexico border.

Like the previous year's appropriations, Congress did not enact legislation providing FY2004 funds for EPA before the beginning of the new fiscal year; thus EPA programs were covered by a series of continuing resolutions (CRs). The last of these CRs (P.L. 108-135) extended FY2003 funding levels through January 31, 2004. On December 8, 2003, the House passed legislation providing full-year funding for EPA and other agencies that lacked enacted appropriations (H.R. 2673). The conference report on this bill (H.Rept. 108-401) provided $1.34 billion for clean water SRF grants, $845 million for drinking water SRF grants, and $425 million in grants for 520 earmarked grants in listed communities, Alaska Native Villages, and U.S.-Mexico border projects.[19] The Senate approved the conference report on January 22, 2004, and President Bush signed the legislation January 23 (P.L. 108-199).

FY2005

The FY2005 EPA appropriation for water infrastructure funds was the lowest total for these programs since FY1997 (the first year in which Congress provided both clean water and drinking water SRF capitalization grants, as well as earmarked project grants). The decline was due primarily to a reduction in funding for the clean water SRF program from an average of $1.35 billion since FY1998 to $1.09 billion.

President Bush's FY2005 budget, presented February 2, 2004, requested a total of $3.0 billion for water infrastructure assistance and state environmental program grants. It included $850 million for clean water SRF grants, $850 million for drinking water SRF grants, $94 million for priority projects (primarily in Alaska Native Villages and along the U.S.-Mexico border), and $1.25 billion for categorical grants. As in recent budgets, the Administration requested no funds for congressionally earmarked project grants. Anticipating that critics likely would focus on the clean water SRF request ($492 million below the FY2004 level), in its budget documents the Administration said that the request included funding for the clean water SRF at $850 million annually through 2011, which, together with loan repayments, state matches and other funding sources, would result in a long-term average revolving level of $3.4 billion. Likewise, the budget anticipated funding the drinking water SRF program at the same $850 million annually through 2011, resulting in a long-term average revolving level of $1.2 billion.

House and Senate Appropriations committees began review of the EPA budget request in March. On September 9, 2004, the House Appropriations Committee reported FY2005 funding for EPA in a bill that included the Administration's requested level of $850 million for clean water SRF grants, $850 million for drinking water SRF grants, and earmarked grants for priority water infrastructure projects totaling $393.4 million (H.R. 5041, H.Rept. 108-674). On September 21, the Senate Appropriations Committee reported its version of this bill (S. 2825, S.Rept. 108-353), which included $1.35 billion for clean water SRF grants, $850 million for drinking water SRF grants, and $217 million for earmarked project grants.

Final action on the FY2005 appropriation did not occur before the start of the fiscal year. On November 20, the House and Senate passed H.R. 4818 (H.Rept. 108-792), the Consolidated Appropriations Act, 2005, an omnibus appropriations bill comprising nine appropriations measures, including funding for EPA. The bill provided total funding for EPA of $8.1 billion,[20] a decrease from the $8.4 billion approved in FY2004, but $340 million more than was requested by the President in February. One of the most controversial items in the final bill was a $251 million decrease for clean water SRF grants from the FY2004 level, although the $1.09 billion total is $241 million more than in the President's budget. The final measure also included $843 million for drinking water SRF capitalization grants; $401.7 million for 669 earmarked grants in listed communities, Alaska Native villages, and U.S.-Mexico border projects; and $1.14 billion for categorical state grants. The $2.34 billion total for water infrastructure programs and projects was $542 million more than was requested by the President, but $276 million less than Congress appropriated for FY2004. President Bush signed the legislation December 8, 2004 (P.L. 108-447).

FY2006

The FY2006 appropriation for water infrastructure funds marked the second consecutive year in which Congress appropriated less funding for these programs, providing lower levels both for clean water SRF capitalization grants and for earmarked project grants than in FY2005.

President Bush presented the FY2006 budget request in February 2005. Overall for EPA, it sought 5.6% less than Congress had appropriated for FY2005. The Administration's deepest cuts affecting EPA were proposed for the STAG account. The budget requested $730 million for clean water SRF grants (33% below FY2005 appropriated funding and 45.6% below the FY2004 level), $850 million for drinking water SRF grants (a slight increase from the FY2005 level), $69 million for priority projects (primarily in Alaska Native Villages and along the U.S.-Mexico border), and $1.2 billion for state categorical grants. As in previous years, the Administration requested no funds for congressionally earmarked water infrastructure projects Advocates for the SRF programs (especially state and local government officials) contended that cuts to the clean water program would impair their ability to carry out needed municipal wastewater treatment plant improvement projects. Administration officials responded that the proposed SRF reductions for FY2006 were because Congress had boosted funds above the FY2005 request level. These officials said that the Administration plans to invest $6.8 billion in the clean water SRF program between FY2004-FY2011, after which federal funding is expected to end, and the state SRFs are expected to have an annual revolving level of $3.4 billion. If Congress appropriates more than is requested in any given year (as occurred in FY2005), they said, that target will be met sooner, leading to reduced requests for the SRF in subsequent years until a planned phaseout in FY2011.

On May 19, 2005, the House passed H.R. 2361, providing FY2006 funding for EPA. As passed, it provided $850 million for clean water SRF grants ($120 million more than the President's request), $850 million for drinking water SRF grants, and $269 million for earmarked water infrastructure grants. During debate, the House rejected two amendments to increase clean water SRF funding. On June 29, the Senate passed its version of H.R. 2361, providing $1.1 billion for clean water SRF grants, $850 million for drinking water SRF grants, and $290 million for earmarked project grants. The House bill required that $100 million of the SRF funding come from balances from expired contracts, grants, and interagency agreements from various EPA appropriation accounts. The Senate bill, in contrast, called for a $58 million rescission of unobligated amounts associated with grants, contracts, and interagency agreements in various accounts, but did not specify that such monies go to SRF funding.

Conferees resolved differences between the bills (H.Rept. 109-188), and the House and Senate approved the measure in July; the President signed it into law on August 2 (P.L. 109-54). As enacted, the bill provided $900 million for clean water SRF grants; $850 million for drinking water SRF grants; $285 million for 259 earmarked grants in listed communities, Alaska Native villages, and along the U.S.- Mexico border; and $1.13 billion for categorical state grants. The final bill required a $80 million rescission from expired grants, contracts, and interagency agreements in various EPA accounts (not just the STAG account) not obligated by September 1, 2006. It did not direct the rescinded funds to be applied to the clean water SRF, as proposed by the House. The $2.03 billion total in the bill for EPA water infrastructure programs and projects was $386 million more than was requested by the President, but $301 million less than Congress appropriated for FY2005.

However, the funding amounts specified in P.L. 109-54 were reduced slightly. First, a provision of P.L. 109-54, Section 439, mandated an across-the-board rescission of 0.476% for any discretionary appropriation in that bill. Second, in December 2005 Congress enacted P.L. 109-148, the FY2006 Department of Defense Appropriations Act, and Section 3801 of that bill mandated a 1% across-the-board rescission for discretionary accounts in any FY2006 appropriation act (except for discretionary authority of the Department of Veterans Affairs). As a result of these two rescissions, the final levels for the STAG account were: $887 million for clean water SRF grants; $838 million for drinking water SRF grants; $281 million for 259 earmarked grants in listed communities, Alaska Native villages, and along the U.S.-Mexico border; and $1.11 billion for categorical state grants. FY2006 EPA water infrastructure programs and projects thus total $2.0 billion.

On October 28, President Bush requested that Congress rescind $2.3 billion from 55 "lower-priority federal programs and excess funds," including $166 million from clean water SRF monies. In the end, Congress did not endorse the specific request to reduce clean water SRF appropriations. The two rescissions resulting from P.L. 109-54 and P.L. 109-148 totaled a $13.2 million reduction from the $900 million specified in the EPA appropriations act.

FY2007

President Bush presented the Administration's FY2007 budget request in February 2006, asking Congress to appropriate $1.570 billion for EPA's water infrastructure programs. The FY2007 request sought $687.6 million for clean water SRF grants, $841.5 million for drinking water SRF grants, and $40.6 million for special projects in Alaska Native Villages, Puerto Rico, and along the U.S.-Mexico border. When the 109th Congress adjourned in December 2006, it had not completed action on appropriations legislation to fund EPA (or on nine other appropriations bills covering the majority of domestic discretionary agencies and departments) for the fiscal year that began October

1, 2006, thus carrying over this legislative activity into the 110th Congress. In December 2006, Congress enacted a continuing resolution, P.L. 109-383 (the third such continuing resolution since the start of the fiscal year on October 1), providing funds for EPA and the other affected agencies and departments until February 15, 2007.

The President's FY2007 budget request for clean water SRF capitalization grants was 22% less than the FY2006 appropriation for these grants and 37% below the FY2005 funding level. The request for drinking water SRF grants was essentially the same as in recent years ($4 million more than FY2006, $1.7 million less than FY2005). As in recent budgets, the Administration proposed no funding for congressionally designated water infrastructure grants, but, as noted above, it did seek a total of $40.6 million for Administration priority projects. Advocates of the clean water SRF program (especially state and local government officials) again contended, as they have for several recent years, that the cuts would impair their ability to carry out needed municipal wastewater treatment plant improvement projects. Administration officials responded that cuts for the clean water SRF in FY2007 were necessary because Congress boosted funds above the requested level in FY2005 and FY2006.

On May 18, 2006, the House passed H.R. 5386 (H.Rept. 109-465), providing the requested level of $687.6 million for clean water SRF grants and $841.5 million for drinking water SRF grants. The Senate Appropriations Committee approved the same funding levels for these grant programs when it reported H.R. 5386 on June 29 (S.Rept. 109-275), but the Senate did not act on this measure before the 109th Congress adjourned in December. Before adjournment, Congress enacted a continuing resolution (CR), P.L. 109-383 (the third such CR since the start of the fiscal year on October 1), providing funds for EPA and the other affected agencies and departments until February 15, 2007. Funding levels provided under this CR followed a "lowest level" concept for individual programs, that is, programs were funded at the lowest level under either House-passed FY2007 appropriations, Senatepassed appropriations, or the FY2006 funding. For clean water SRF grants, the resulting appropriation through mid-February was $687.6 million, as in House-passed H.R. 5386. For drinking water SRF grants, the appropriation level through mid-February was $837.5 million, the FY2006-enacted level. The CR included funds for congressionally earmarked water infrastructure project grants totaling $200 million, as in House-passed H.R. 5386.

Returning to these issues in 2007, in mid-February, Congress passed H.J.Res. 20, a continuing appropriations resolution that provides funding for

EPA and the other affected agencies through the end of FY2007. As passed, this full-year resolution held most programs and activities at their FY2006 appropriated levels. However, clean water SRF capitalization grants were one of the few programs that received a funding increase under the resolution: these grants received $1.08 billion ($197 million more than in FY2006, and $396 million more than the President requested for FY2007). The resolution further prohibited project grants for congressional earmarks, but not for special project grants requested in the President's budget. The action to ban earmarks in FY2007 occurred when leaders in the 110th Congress sought to finish up appropriations actions that were unresolved at the end of the 109th Congress, and at the same time the newly elected Congress moved to adopt rules and procedures to reform the congressional earmarking process for the future. (Water infrastructure project earmarks totaled $281 million in EPA's FY2006 appropriation.) President Bush signed H.J.Res. 20 on February 15 (P.L. 110-5).

The final FY2007 amounts provided in P.L. 110-5 were

- $1.084 billion for clean water SRF capitalization grants,
- $837.5 million for drinking water SRF capitalization grants,
- $83.75 million for Alaskan Native Village and U.S.-Mexico Border project grants requested by the Administration, and
- $1.11 billion for categorical state grants.

FY2008

The President's FY2008 budget request was presented to Congress on February 5, 2007, before finalization of the FY2007 appropriations. The budget sought $687.6 million for clean water SRF grants, the same amount requested for FY2007; $842.2 million for drinking water SRF grants; $25.5 million for special project grants for Alaskan Native Villages and the U.S.-Mexico Border region; and $1.065 billion for categorical state grants.

In June 2007, the House passed H.R. 2643, providing FY2008 appropriations for EPA. This bill included $1.125 billion for clean water SRF grants, $842.2 million for drinking water SRF grants, plus $175.5 million for 143 congressionally designated water infrastructure project grants. The Senate Appropriations Committee approved companion legislation (S. 1696) that similarly included higher funding levels for several water quality programs. The Senate committee's bill provided less funding for clean water SRF grants than the House bill ($887 million), the same amount for drinking water SRF grants, and slightly more for congressionally designated water infrastructure project grants ($180 million). The Senate did not take up S. 1696.

By October 1, the start of FY2008, Congress had not enacted any appropriations bills for FY2008, and Congress enacted several short-term continuing appropriations resolutions to temporarily fund EPA and other government agencies until final agreement, which occurred in December 2007. Full-year funding for EPA's water infrastructure programs was included in the Consolidated Appropriations Act for FY2008 (Division F, Title II), signed by the President December 26, 2007 (P.L. 110-161).

The final FY2008 amounts[21] provided in this legislation were

- $689.1 million for clean water SRF capitalization grants ($1.5 million more than requested by the Administration),
- $829.0 million for drinking water SRF capitalization grants ($13.2 million less than requested),
- $177.2 million for 282 earmarked grants in listed communities, Alaska Native villages, and U.S.-Mexico border projects ($151.7 million more than requested), and
- $1.078 billion for categorical state grants ($13.3 million more than requested).

FY2009

The President's FY2009 budget request was presented to Congress on February 6, 2008. The budget sought $555 million for clean water SRF grants, $134 million less than Congress appropriated for FY2008; $842.2 million for drinking water SRF grants, $13 million more than was appropriated for FY2008; $25.5 million for special project grants for Alaskan Native Villages and the U.S.-Mexico Border region, $18.8 million less than was appropriated for FY2008; and $1.057 billion for categorical state grants. As in past years, the budget requested no funds for other earmarked grants.

Final congressional action on most appropriations bills, including funding for EPA, is not expected to occur until after the November 2008 election.

End Notes

[1] U.S. Environmental Protection Agency, Office of Wastewater Management, *Clean Watersheds Needs Survey 2000, Report to Congress,* August 2003, EPA-832-R-03-001, 85 p.

[2] For information, see CRS Report RS22037, *Drinking Water State Revolving Fund: Overview and Issues*, by Mary Tiemann.

[3] U.S. Environmental Protection Agency, *Drinking Water Infrastructure Needs Survey,Third Report to Congress,* June 2005,EPA-816-R-05-001, 71 p.

[4] Prior to the 109th Congress, EPA appropriations were included in legislation funding the Department of Veterans Affairs, Department of Housing and Urban Development, and Independent Agencies (VA/HUD). In January 2005, House and Senate Appropriations Committees reorganized, and jurisdiction over funding for EPA and several other entities was moved to the Appropriations subcommittees covering Interior and Related Agencies.

[5] For additional information, see CRS Report RL32201, *Water Infrastructure Project Earmarks in EPA Appropriations: Trends and Policy Implications*, by Claudia Copeland.

[6] In the 104th Congress, the House passed a comprehensive CWA reauthorization bill that included provisions concerning water infrastructure (H.R. 961), but provisions in it that addressed regulatory relief and similar issues were controversial, and no further action occurred. In the 107th, 108th, and 109th Congresses, House and Senate committees considered legislation to reauthorize water infrastructure financing programs, but no bill was enacted. In the 110th Congress, the House has passed a reauthorization bill (H.R. 720), but no action has occurred in the Senate.

[7] H.Rept. 104-715, accompanying H.R. 4624, 103rd Cong., 2d sess., p. 42.

[8] H.Rept. 104-812, accompanying H.R. 3666, 104th Cong., 1st sess., p. 74.

[9] H.Rept. 105-175, accompanying H.R. 2158, 105th Cong., 1st sess., p. 69; S.Rept. 105-216, accompanying S. 2168, 105th Cong., 2nd sess., p. 82.

[10] H.Rept. 106-379, accompanying H.R. 2684, 106th Cong., 1st sess., p. 141.

[11] Section 104(b)(3) grants have been used to support a variety of special studies and projects allowing states and localities to demonstrate innovative approaches to implementing the core water quality program.

[12] *Colonias* are unincorporated areas outside city boundaries along the U.S.-Mexico border. Most lack adequate public utilities, especially water and wastewater services.

[13] The 1987 Clean Water Act amendments authorized federal grants to assist states in implementing programs to manage water pollution from nonpoint sources such as farm and urban areas, construction, forestry, and mining sites. Because of competing demands for funding, it had been difficult for Congress to fund this grant program and other water quality initiatives in the 1987 Act. Appropriators did fund Section 319 grants in EPA's general program management account (abatement, control and compliance) in FY1990, FY1991, and FY1992, but well below authorized levels. In the FY1993 Act, appropriators moved funding into the SRF/construction grants account, thereby providing a degree of protection from competing priorities.

[14] H.Rept. 104-384, accompanying H.R. 2099, 104th Cong., 1st sess., in Congressional Record, vol. 141, no. 193, daily ed. December 6, 1995, p. H 14132. This was the second conference report on this bill; a previous agreement, reflected in H.Rept. 104-353, was rejected by the House on November 29. However, amounts in the State and Tribal Assistance Grants account were the same in both versions.

[15] The conference report on H.R. 3019 (H.Rept. 104-537) references the conference report on the vetoed H.R. 2099, making the two reports together the full statement of the conference committee regarding EPA funding and the State and Tribal Assistance Grants account.

[16] S.Rept. 105-53, accompanying S. 1034, 105th Cong., 1st sess., p. 71.

[17] Office of Management and Budget. "Cancellation Pursuant to Line Item Veto Act," 62 Federal Register 59768, November 4, 1997. The President also cancelled funding for two other

projects in the EPA portion of the bill, a water and wastewater training institute in Alabama and a solar aquatic wastewater treatment plant in Vermont. These projects were funded under a separate EPA account in the bill, the environmental programs and management account.

[18] H.J.Res. 2 included an across-the-board 0.65% reduction to accounts funded by the legislation, and to each program, project, and activity within an account. This reduction is reflected in amounts described here.

[19] H.R. 2673 included an across-the-board 0.59% reduction to accounts funded by the legislation, and to each program, project, and activity within an account. This reduction is reflected in amounts described here.

[20] H.R. 4818 included an across-the-board 0.80% reduction to accounts funded by the legislation, and to each program, project, and activity within an account. This reduction is reflected in amounts described here.

[21] Finals amounts shown here reflect a 1.56% across-the-board reduction of appropriated amounts for accounts included in Division F of the legislation.

In: Water Infrastructure
Editor: Harald Nystroem
ISBN: 978-1-60692-479-2

Chapter 9

STATEMENT OF MCGINNIS, CHARLES I., COMMITTEE ON ORGANIZING TO MANAGE CONSTRUCTION AND INFRASTRUCTURE IN THE 21ST CENTURY, BUREAU OF RECLAMATION BOARD ON INFRASTRUCTURE AND THE CONSTRUCTED ENVIRONMENT DIVISION ON ENGINEERING AND PHYSICAL SCIENCES, THE NATIONAL ACADEMIES

Charles I. McGinnis

Good afternoon, Mr. Chairman and members of the Committee. My name is Charles I. McGinnis. I am a retired Major General and former director of civil works for the U.S. Army Corps of Engineers. I served on the National Research Council Committee which authored the report, *Managing Construction and Infrastructure in the 21st Century Bureau of Reclamation.* The National Research Council is the operating arm of the National Academy of Sciences, National Academy of Engineering, and the Institute of Medicine of the National Academies, and its goal is to provide elected leaders, policy makers, and the public with expert advice based on sound scientific evidence.

I appreciate the opportunity to be here to discuss our report, which was published in 2006. The report contained recommendations on a broad range of issues, including organizational structure, policy development, project management, acquisition and contracting, and stakeholder relationships. Today I will focus on those issues and recommendations that pertain to the management, operation, and maintenance of Reclamation's aging infrastructure, and provide a brief summary of the Bureau's response to the report. My written testimony, which has been submitted for the record, includes an appendix with additional information from the study.

STUDY BACKGROUND

The study committee was asked by the Department of the Interior to advise Reclamation and the department on the "appropriate organizational, management, and resource configurations to meet its construction, maintenance, and infrastructure requirements for its missions of the 21st century." The committee was comprised of 12 experts from the public and private sectors and academia. We met as a whole four times from February to August 2005 and conducted small-group site visits to offices and projects in each of the five Reclamation regions. We received briefings from and had discussions with Reclamation representatives, Reclamation's customers and other stakeholders, and representatives of organizations with missions similar to Reclamation's, including the U.S. Army Corps of Engineers, the Tennessee Valley Authority, and the California Department of Water Resources.

ASSET MANAGEMENT ISSUES

Since its establishment in 1902, Reclamation has constructed more than 500 dams and hydropower plants, and more than 300 related structures including pumping plants, fish protection facilities, and buildings. At this time, however, relatively few large new projects are planned. As a consequence, Reclamation's focus and workload have shifted from building new facilities to operating, maintaining, repairing, and modernizing existing ones, and to evaluating dam safety, providing for dam security, and addressing environmental issues.

This transition brings with it significant changes in the workload and in the responsibilities, duties, and activities of the workforce. Reclamation's current work is dominated by two categories of tasks: (1) the operation, maintenance and rehabilitation of existing structures and systems; and (2) the creation and brokering of agreements among a variety of groups and interests affected by the management of water resources.

Although its activities have changed, Reclamation's mission continues to be the effective management of power and water resources in ways that protect the health, safety, and welfare of the American public and are environmentally and economically sound. Achieving these objectives will depend on Reclamation's ability to effectively manage a number of constraints and realities. These include:

- *Aging infrastructure.* Many of Reclamation's dams, power plants, and related infrastructure are more than 50 years old, and some are almost 100 years old. Most embody out-of-date design, engineering practices, and materials. Their age increases their maintenance requirements as the structures and equipment reach or pass their design lifetimes, and wear out through daily use.
- *Transferred works.* Some facilities are owned by Reclamation but operated and maintained by users such as water districts. These "transferred works" are generally irrigation-system-related facilities, including smaller dams, dikes, pumping plants, and canals. The resources and sophistication of the water districts that operate and maintain transferred works vary. Although some districts are willing and able to perform a larger role, others have fewer resources. Some water customers already find full payment for operation and maintenance activities difficult, and major repairs and modernization needs, if included in the operations and maintenance budget, impose an even greater financial burden that cannot be met under the current repayment requirements.
- *Increasing competition for declining resources.* Although water availability is declining in many parts of the West, existing water users continue to demand reliable systems to provide as much water as they have used historically. Additional demands are posed by environmental requirements and by increases in population and industry.
- *Increased regulatory requirements.* Water rights regulations, Endangered Species Act (ESA) requirements, environmental impact

assessment (EIA) requirements, and expectations for increased openness and public involvement in decision making place additional demands on Reclamation's project managers, operators, and decision makers.

- *Security*. Security reviews and ongoing security management add to the workload at many of the larger facilities, including those facilities designated national critical infrastructure.

Considering these trends and changes, the study committee made recommendations for Reclamation to develop the appropriate organizational, managerial, and resource configurations to meet its construction, maintenance, and infrastructure requirements for its missions of the twenty-first century. I should point out that our recommendations were purposely general in nature as the study committee believed that the specifics could be best developed internally where more detailed knowledge resides.

PLANNING FOR ASSET SUSTAINMENT

The Bureau of Reclamation has a decentralized management structure in which each of its five regions is responsible for sustaining a significant portfolio of facilities. The regions have different organizational structures, capabilities, and workloads. The regions also have ongoing but different procedures and methods for tracking the maintenance workload and backlog of needs. In power facilities, computerized maintenance management systems are used. Critical maintenance problems receive immediate attention. Less-than-critical needs are prioritized and scheduled as funds become available. At water management assets, needs beyond the scope of normal day-to-day maintenance are tracked through the dam safety information system (DSIS) and replacement, addition, and exceptional maintenance (RAX) lists. The RAX lists are also used to prioritize maintenance needs and funds through the budget formulation process. Budget proposals originate at the area offices, and are then refined and consolidated at the regional and headquarters levels.

The committee observed inconsistencies in the way these processes operate and in how the beneficiaries (primarily water districts) are engaged in decision making and review. Some beneficiaries noted that the rules seem to differ within regions and across regions with respect to who must pay, how much must be paid, and how design and construction activities are carried out. The quality and consistency of assessment and planning documents, except those associated with the larger power facilities, also vary from region to region.

As the owner of facilities, Reclamation headquarters has the responsibility to ensure that its facilities are planned, designed, constructed, and managed with a level of quality that is consistent throughout the Bureau. To demonstrate consistency, Reclamation needs clear, detailed policy directives and standards to enable each of the regions to implement a uniform, structured approach.

Effective planning is the key to the effective operation and maintenance of Reclamation facilities. The committee observed that, in general, the regions will need to evaluate their asset inventory and manage their assets more aggressively over the life cycle and engage in constructive relationships with customers and stakeholders.

In two regions, the committee observed effective processes for planning and executing facility operations and maintenance. The core of these processes consists of 5- and 10-year plans developed to identify out-year funding requirements and to ensure that stakeholders are informed well in advance of future funding requirements, especially for refurbishment.

The operations and maintenance burden for an aging infrastructure will increase, and the financial resources available to Reclamation, its customers, and contractors may not be able to keep up with the increased demand. A number of water districts pointed out to the committee the difficulties resulting from the requirement to reimburse expenditures for operation and maintenance activities within the fiscal year in which they were expended. This is a particular difficulty for some water districts that do not have enough control over cash flow and other factors to do this when operation and maintenance costs increase. Better long-term planning should allow these districts to anticipate such needs. Long-term sustainment will require more innovation and greater efficiency in order to get the job done.

The committee recommended that all regions develop and use 5- and 10-year plans as a stakeholder communications tool and as a roadmap for meeting future requirements. The comprehensive operations and maintenance plans should also serve as the basis for financial management and the development of fair and affordable repayment schedules. The committee also recommended that Reclamation should include its customers in their efforts to address economic constraints by seeking repayment procedures that ease borrowing requirements and extend repayment periods.

Benchmarking and Best Practices

The committee observed extensive efforts and success in benchmarking Reclamation's hydropower activities; however, there appears to be little effort to benchmark the operations and maintenance of water distribution facilities. In the committee's opinion, benchmarking can help improve the efficiency of Reclamation's water management and distribution activities as well as those of the water contractors responsible for transferred works.

In the case of the larger hydroelectric generating facilities, Reclamation uses an independent benchmarking process to determine how its facilities compare to others in terms of costs, reliability, efficiency, and overall maintenance. Such reviews are conducted on an annual basis, and the reports provide useful information to facility managers.

Similar efforts should be made to establish metrics and measure the performance of Reclamation's water management assets. Reclamation regional offices reported the use of some review tools, including annual, periodic, and comprehensive facility reviews, value engineering reviews, and peer review of endangered species recovery programs. The committee was also informed that there are several forums within Reclamation to identify best practices for asset management. However, there seem to be wide differences in the application and dissemination of review tools and best practices across the bureau.

The committee recommended that benchmarking of water distribution and irrigation activities by Reclamation and its contractors should be a regular part of their ongoing activities.

Bureau of Reclamation Response

An important element in the committee's ability to complete its assigned tasks was the support and participation of the bureau. The study committee appreciated the cooperation and support of former Commissioner John Keys III and all of the Reclamation officials who assisted the committee in the review. Before completing our work, we became aware that the commissioner had directed the development of a detailed response to its recommendations. The NRC committee applauded this rapid and enthusiastic response. We were not in a position to provide a detailed analysis at that time, but it appears that Reclamation's response, *Managing for Excellence,* sets forth a plan to address all of the issues identified in the study. Many of the study committee's recommendations will require further analysis by Reclamation personnel, and changes that implement these initiatives may take several years. As noted in the NRC report, Reclamation should seek independent reviews of its assessments and organizational changes. Nevertheless, it appeared that the Bureau had made a good start in implementing the committee's recommendations.

APPENDIX

The report *Managing Construction and Infrastructure in the 21st Century Bureau of Reclamation* contains additional observations and recommendations, as summarized below.

ORGANIZATIONAL STRUCTURE

The study committee recognizes that organizations can and do take on a variety of forms with varying degrees of success. Some will function successfully despite their form, while others will falter even as they deploy the best of theoretical forms. The internal culture and history of an organization play a significant role in determining the appropriate structure and the ultimate outcome. We believe that the organizational structure of Reclamation is basically appropriate for its customer-driven mission to deliver power and water. Nevertheless, we also believe that there are opportunities to improve the construction and management of its facilities and infrastructure, as well as the management, development, and protection of water and related resources in an environmentally sound manner in the interest of the American public.

CENTRALIZED POLICY AND DECENTRALIZED OPERATIONS

To optimize the benefits of decentralization, Reclamation should promulgate policy guidance, directives, standards, and how-to documents that are consistent with the current workload. The commissioner should expedite the preparation of such documents, their distribution, and instructions for their consistent implementation. Reclamation's operations should remain decentralized and guided and restrained by policy but empowered at each level by authority commensurate with assigned responsibility to respond to customer and stakeholder needs. Policies, procedures, and standards should be developed centrally and implemented locally. The design groups in area and project offices should be consolidated in regional offices or regional technical groups to provide a critical mass that will allow optimizing technical competencies and providing efficient service. Technical skills in the area offices should focus on data collection, facility inspection and evaluation, and routine operations and maintenance (O&M).

Technical Service Center and Reclamation Laboratory and Research Activities

The commissioner should undertake an in-depth review and analysis of the TSC to identify the needed core technical competencies, the number of technical personnel, and how the TSC should be structured for maximum efficiency to support the high-level and complex technical needs of Reclamation and its customers. The proper size and composition of the TSC are dependent on multiple factors, some interrelated:

- Forecast workload,
- Type of work anticipated,
- Definition of activities deemed to be inherently governmental,
- Situations where outsourcing may not be practical,
- Particular expertise needed to fulfill the government's oversight and liability roles,
- Personnel turnover factors that could affect the retention of expertise, and
- Needs for maintaining institutional capability.

This assessment and analysis should be undertaken by Reclamation's management and reviewed by an independent panel of experts, including stakeholders.

The workforce should be sized to maintain the critical core competencies and technical leadership, and to increase outsourcing of much of the engineering and laboratory testing work. Alternative means should be developed for funding the staff and operating costs necessary for maintaining core TSC competencies, thereby reducing the proportion of engineering service costs reimbursable by customers.

Reclamation's Research Office and TSC laboratory facilities should be analyzed from the standpoint of which specific research and testing capabilities are required now and anticipated for the future; which of them can be found in other government organizations, academic institutions, or the private sector; which physical components should be retained; and which kinds of staffing are necessary. The assessment should also recognize that too much reliance on outside organizations can deplete an effective engineering capability that, once lost, is not likely to be regained. In making this assessment Reclamation should take into account duplication of facilities at other government agencies, opportunities for collaboration, and the possibility for broader application of numerical modeling of complex problems and systems. Considering that many of the same factors that influence the optimum size and configuration of the TSC engineering services also apply to the research activities and laboratories, Reclamation should consider coordinating the reviews of these two functions.

Outsourcing

Reclamation should establish an agency-wide policy on the appropriate types and proportions of work to be outsourced to the private sector. O&M and other functions at Reclamation-owned facilities, including field data collection, drilling operations, routine engineering, and environmental studies, should be more aggressively outsourced where objectively determined to be feasible and economically beneficial.

Project Management

Reclamation should establish a comprehensive set of directives for structured project management process for managing projects and stakeholder engagement from inception through completion and the beginning of O&M. Reclamation should also give high priority to completing and publishing cost estimating directives and resist pressures to submit projects to Congress with incomplete project planning. Cost estimates that are submitted should be supported by a design concept and planning, environmental assessment, and design development documents that are sufficiently complete to support the estimates.

Reclamation should establish a structured project review process to ensure effective review and oversight from inception through completion of construction and the beginning of O&M. The level of review should be consistent with the cost and inherent risk of the project and include the direct participation of the commissioner or his or her designated representative in oversight of large or high-risk projects. The criteria for review procedures, processes, documentation, and expectations at each phase of the project need to be developed and applied to all projects, including those approved at the regional level.

A training program that incorporates current project management and stakeholder engagement tools should be developed and required for all personnel with project management responsibilities. In addition, project managers should have professional certification and experience commensurate with their responsibilities.

ACQUISITION AND CONTRACTING

Reclamation should establish a procedure and a central repository for examples of contracting approaches and templates that could be applied to the wide array of contracts in use. This repository should be continually maintained and upgraded to allow staff to access lessons learned from use of these instruments.

RELATIONSHIPS WITH SPONSORS AND STAKEHOLDERS

Making information readily available about processes and practices, both in general and for specific projects and activities, should be a Reclamation priority. Successful practices, such as those used in the Lower Colorado Dams Office, should be analyzed and the lessons learned should be transferred, where practical, throughout the bureau.

WORKFORCE AND HUMAN RESOURCES

Reclamation should do an analysis of the competencies required for its personnel to oversee and provide contract administration for outsourced activities. Training programs should ensure that those undertaking the functions of the contracting officer's technical representative are equipped to provide the appropriate oversight to ensure that Reclamation needs continue to be met as mission execution is transferred.

In light of the large number of retirements projected over the next few years and the potential loss of institutional memory inherent in these retirements, a formal review should be conducted to determine what level of core capability should be maintained to ensure that Reclamation remains an effective and informed buyer of contracted services. Reclamation should recruit, train, and nurture personnel who have the skills needed to manage processes involving technical capabilities as well as communications and collaborative processes. Collaborative competencies should be systematically related to job categories and the processes of hiring, training, evaluating the performance of, and promoting employees. Reclamation should facilitate development of the skills needed for succeeding at socially and politically complex tasks by adapting and adopting a small-wins approach to organizing employee efforts and taking advantage of the opportunities to celebrate and build on successes.

CHAPTER SOURCES

The following chapters have been previously published:

Chapter 1 – This is an edited, reformatted and augmented version of a Congressional Research Service publication, Report Order Code RL33108, updated March 25, 2008.

Chapter 2 - This is an edited, reformatted and augmented version of a Congressional Research Service publication, Report Order Code RL34466, updated April 30, 2008.

Chapter 3 – These remarks were delivered as testimony given on April 17, 2008. Wesley Randal "Randy" Reed, Co-chair St. Mary Rehabilitation Working Group, Senate Committee on Energy and Natural Resources.

Chapter 4 – These remarks were delivered as testimony given on April 17, 2008. Tony Willardson, Deputy Director, Western States Water Council before the Subcommittee on Water and Power of the Energy and Natural Resources Committee, United States Senate.

Chapter 5 - These remarks were delivered as testimony given on April 17, 2008. Thomas F. Donnelly, Executive Vice President, National Water Resources Association before the Subcommittee on Water and Power of the Energy and Natural Resources Committee, United States Senate.

Chapter 6 - These remarks were delivered as testimony given on April 17, 2008. Robert W. Johnson, Commissioner, Bureau of Reclamation, United States Department of the Interior before the Subcommittee on Water and Power of the Energy and Natural Resources Committee, United States Senate.

Chapter 7 - These remarks were delivered as testimony given on April 17, 2008. Mayor Todd Cutler, City of Fernley, Nevada before the Subcommittee on Water and Power of the Energy and Natural Resources Committee.

Chapter 8 - This is an edited, reformatted and augmented version of a Congressional Research Service publication, Report Order Code 96-647, updated August 19, 2008.

Chapter 9 - These remarks were delivered as testimony given on April 17, 2008. Charles I. McGinnis, United States Army Corps of Engineers (retired), Member, Committee on Organizing to Manage Construction and Infrastructure in the 21st Century Bureau of Reclamation before the Subcommittee on Water and Power of the Energy and Natural Resources Committee.

INDEX

A

abatement, 114
academic, 125
accounting, 88
administration, 93, 127
administrators, 51, 82
affirmative action, 53
afternoon, 117
age, vii, viii, ix, 1, 5, 10, 14, 18, 23, 24, 28, 32, 33, 34, 68, 72, 119
aging, vii, viii, ix, 3, 6, 17, 23, 24, 25, 28, 32, 33, 34, 37, 38, 39, 40, 47, 49, 52, 53, 57, 59, 63, 65, 67, 68, 118, 121
agricultural, 30, 37, 38, 41, 43, 53, 61, 62, 63, 66, 78
agriculture, 12, 39, 40, 43, 49
aid, vii, ix, 26, 32, 35, 42, 56, 75, 76, 78, 88
air, 80, 93, 95
air force, 11
air quality, 80
Alaskan Native, 94, 95, 96, 98, 99, 100, 102, 104, 106, 112, 113
alfalfa, 40, 43
alternatives, 45, 46, 53, 54, 56
amendments, x, 14, 76, 77, 78, 79, 82, 83, 85, 89, 95, 96, 97, 109, 114
appendix, 118
application, 32, 84, 122, 125
appropriations bills, 31, 82, 83, 84, 93, 94, 104, 110, 113
appropriations committee, 48, 84, 96, 99, 105, 108, 111, 112, 114
aquifers, 63
arid, 24, 62
army, ix, x, 3, 11, 13, 20, 21, 24, 25, 29, 35, 68, 117, 118, 130
Army Corps of Engineers, ix, x, 3, 13, 20, 21, 24, 25, 35, 68, 117, 118, 130
assessment, 12, 13, 25, 32, 56, 60, 120, 124, 125
assets, 24, 25, 40, 49, 67, 68, 69, 70, 98, 120, 121, 122
attacks, 2, 104
authority, 9, 12, 20, 28, 29, 47, 55, 61, 81, 83, 98, 101, 110, 124
availability, 13, 32, 119

B

balanced budget, 97
banking committee, 52
barley, 40

barrier, 3, 43, 66
benchmark, 122
benefits, vii, 1, 2, 5, 39, 40, 43, 44, 46, 49, 52, 56, 60, 61, 62, 123
best practice, 122
block grants, 94
bonding, 15, 16, 47
bonds, 55
borrowing, 54, 121
budget deficit, 87
budgetary resources, 87
buildings, 118
Bureau of Land Management, 11, 48
Bureau of Reclamation, v, vi, vii, viii, ix, 1, 8, 11, 14, 20, 23, 24, 34, 35, 37, 38, 39, 40, 41, 42, 48, 49, 51, 52, 53, 55, 59, 60, 61, 62, 65, 71, 117, 120, 122, 123, 129, 130
Bush Administration, 103
buyer, 127

C

canals, 27, 29, 33, 67, 119
capital expenditure, 39, 49, 79
caps, 33
cargo, 31
cash flow, 121
catastrophes, viii, 1
category a, 61
cattle, 43
certification, 126
citizens, 48, 72
classification, 5, 6
clean air, 95
Clean Water Act, vii, ix, x, 75, 76, 79, 81, 82, 83, 89, 101, 114
cleanup, 97
Clinton Administration, 90
coal mine, 12
collaboration, 125
collateral, 26
committee on appropriations, 20, 35
commodity, 56
Commodity Credit Corporation, 12
communities, x, 15, 39, 40, 41, 44, 48, 49, 52, 53, 56, 62, 67, 71, 72, 73, 76, 78, 79, 81, 82, 83, 84, 90, 91, 92, 94, 97, 98, 100, 102, 105, 106, 107, 108, 109, 110, 113
competition, 119
complexity, 33
compliance, 48, 114
components, 5, 9, 28, 34, 38, 41, 42, 44, 125
composition, 124
concrete, 27, 29, 42
configuration, 125
conflict, 28, 33
congressional record, 114
conservation, 59
Consolidated Appropriations Act, 101, 108, 113
constraints, 119, 121
consultants, 12
contaminants, 78
contractors, 121, 122
contracts, 12, 39, 42, 43, 44, 49, 56, 109, 110, 126
control, 2, 3, 4, 12, 16, 17, 25, 53, 66, 77, 114, 121
cost-effective, 54
costs, vii, x, 1, 2, 12, 14, 15, 27, 28, 29, 30, 33, 39, 49, 60, 62, 67, 68, 69, 76, 77, 79, 83, 121, 122, 124
cost-sharing, 30
covering, 110, 114
cracking, 10, 29
credit, 55
creeks, 42
critical infrastructure, viii, 2, 120
CRS, 1, 23, 34, 35, 75, 99, 114
crude oil, 67
culture, 123
customers, 44, 54, 67, 68, 69, 70, 118, 119, 121, 124, 125
cutters, 77

D

data collection, 48, 124, 125
database, 3, 19
deaths, 5
decentralization, 123
decision makers, 120
decisions, 3, 25, 26, 53, 69
defense, 106
deficit, 87
definition, 9
degradation, 42
delivery, 38, 39, 41, 42, 49, 54, 62
Demonstration Project, 89
Department of Agriculture, 4, 11, 30
Department of Defense, 4, 11, 13, 30, 110
Department of Energy, 4, 11
Department of Homeland Security, 17
Department of Housing and Urban Development, 114
Department of Interior, 38, 47
Department of State, 11
Department of the Interior, vi, ix, 4, 10, 11, 14, 24, 65, 75, 79, 118, 129
desert, 71
directives, 121, 123, 125
disaster, 72, 73
discretionary, 12, 17, 98, 101, 110
disputes, 105
distribution, 25, 63, 87, 96, 102, 122, 123
division, 31
domestic policy, 77
doors, 73
drinking, x, 2, 40, 75, 76, 78, 79, 81, 83, 85, 91, 92, 93, 94, 95, 96, 97, 98, 99, 100, 101, 102, 103, 104, 105, 106, 107, 108, 109, 110, 111, 112, 113
drinking water, x, 40, 75, 76, 78, 79, 81, 83, 85, 91, 92, 93, 94, 95, 96, 97, 98, 99, 100, 101, 102, 103, 104, 105, 106, 107, 108, 109, 110, 111, 112, 113
drought, 43, 53
duplication, 125
duties, 119

E

earthquake, 18
economic development, 62
economies of scale, 56
election, 113
electricity, 31
eligibility criteria, 32
employees, 66, 67, 127
empowered, 124
encouragement, 9
Endangered Species Act, 119
energy, 24, 54, 55, 62, 67
Energy and Water Development, 27, 31, 33
engagement, 125, 126
environment, 2, 5, 50, 56, 94
environmental impact, 42, 46, 119
environmental issues, 118
environmental protection, 52, 55, 93
Environmental Protection Agency (EPA), vi, vii, ix, x, 75, 76, 77, 78, 79, 80, 82, 83, 84, 86, 88, 93, 94, 95, 97, 98, 99, 100, 101, 102, 103, 104, 105, 106, 107, 108, 109, 110, 111, 112, 113, 114, 115
estimating, 125
evacuation, 72
exclusion, 33
execution, 127
expenditures, vii, viii, ix, 23, 24, 39, 49, 54, 79, 121
expertise, 66, 124

F

failure, vii, viii, 1, 2, 3, 5, 7, 10, 17, 19, 38, 45, 46, 62, 66, 69, 72
family, 40
Farm Bill, 20

Farm Security and Rural Investment Act, 20, 21
farmers, 67
farmland, 24, 62, 67
farms, 43
fears, 72
Federal Emergency Management Agency (FEMA), 5, 8, 9, 10, 17, 19, 20, 35, 72
Federal Energy Regulatory Commission (FERC), viii, 2, 4, 8, 12, 20
federal funds, ix, 24, 30, 46, 47, 48, 49
federal government, viii, 1, 3, 4, 7, 12, 14, 17, 18, 26, 38, 39, 40, 47, 49, 60, 62, 72, 87
federal grants, x, 76, 78, 82, 114
Federal Register, 114
Federal Water Pollution Control Act, 76
fees, 15, 82
feet, 3, 24, 29, 41, 42, 43, 44, 67, 72
fiber, 53
finance, viii, 1, 54, 55, 79
financial resources, viii, 2, 56, 60, 121
financial support, 48
financing, x, 51, 55, 56, 62, 69, 76, 77, 78, 82, 91, 99, 114
fire, 3
fiscal policy, 61
fish, 3, 43, 53, 63, 118
Fish and Wildlife Service, 11, 45
fisheries, 63
fishing, 40, 44
flexibility, 77, 78, 84, 93, 101, 102
flood, 2, 3, 4, 12, 16, 17, 18, 31, 42, 53, 62, 68, 72
flow, 43, 44, 66, 87, 121
food, 44, 53
Forest Service, 11
forestry, 114
fruits, 24
fuel, 62

G

gas, 87
generation, 31, 37, 62
Glacier National Park, 42
goals, 46, 53, 54
goods and services, 56
government, iv, viii, x, 1, 2, 3, 4, 7, 12, 14, 15, 16, 17, 18, 20, 26, 38, 39, 40, 47, 49, 56, 60, 62, 67, 69, 72, 76, 77, 79, 87, 93, 101, 109, 111, 113, 124, 125
governors, 51
grass, 48
grassroots, 59
greed, 85, 87, 88
groundwater, 71, 84, 96, 98
groups, 3, 15, 82, 101, 104, 106, 119, 124
growth, 52, 53, 55, 67
guidance, 10, 28, 123
guidelines, 7, 8, 25, 34, 84
guiding principles, 54

H

habitat, 2, 43, 44, 45, 53, 63
hands, 53
harm, 5
hazards, 13, 28, 32
health, 56, 78, 79, 119
hearing, 60
height, 3
high-level, 124
high-risk, 126
hiring, 127
Homeland Security, 8, 17, 19, 20, 35
house, 82, 84, 86, 94, 95, 96, 97, 98, 100, 101, 102, 103, 104, 105, 106, 107, 108, 109, 111, 112, 114
House Appropriations Committee, 105, 108
housing, 72
Housing and Urban Development (HUD), 114

human, 3, 6, 9
hunting, 40
Hurricane Katrina, 2, 5, 18
hydroelectric power, 2, 24, 31, 67
hydrologic, 14, 28
hydropower, 4, 12, 31, 118, 122

I

ice, 72
identification, 14, 26, 28
impact assessment, 120
implementation, 29, 34, 39, 40, 49, 87, 91, 123
incentives, 56
Incidents, 19
inclusion, x, 6, 19, 75, 81
Independent Agencies, 114
Indian, 11, 29, 38, 39, 40, 43, 46, 47, 49, 62, 94
indices, 24
individual action, 31
industrial, 2, 43, 53, 56, 78, 99
industry, 12, 39, 40, 49, 119
inertia, 38, 47
information exchange, 8
information technology, 9
injury, iv
innovation, 121
inspections, vii, viii, 1, 2, 7, 8, 9, 10, 11, 12, 14, 15, 18, 19, 24, 27, 66, 73, 124
Inspector General, 20
inspectors, viii, 2, 9
instability, 29
institutions, 125
instruments, 126
insurance, 55, 73
integrity, 3, 6, 18, 28, 70, 77, 101
interest groups, 106
interstate, 52, 54
inventories, 30
investment, 52, 53, 79, 90
irrigation, vii, viii, 2, 3, 16, 23, 24, 25, 27, 39, 41, 42, 43, 49, 60, 61, 62, 67, 68, 71, 119, 122

J

jurisdiction, viii, 2, 52, 114

K

Katrina, 2, 5, 18
killing, 19

L

land, 19, 43, 54, 56, 78
language, 83, 84, 86, 90, 92, 93, 94, 95, 96, 98, 99, 101, 104, 105, 106
large-scale, 40, 43
law, 4, 7, 35, 52, 61, 67, 72, 87, 89, 100, 105, 109
law suits, 72
lawyers, 72
leadership, 38, 39, 46, 47, 48, 124
leakage, 29
leaks, 42
legislation, ix, 6, 8, 14, 17, 27, 30, 33, 61, 73, 75, 79, 81, 83, 85, 86, 88, 91, 92, 93, 94, 95, 98, 103, 104, 106, 107, 108, 110, 112, 113, 114, 115
lenders, 26
levees, viii, 2, 3
lien, 15
life cycle, 121
lifespan, 6
likelihood, 5
limitation, 9
loan guarantees, 55, 61, 70
loans, x, 15, 16, 76, 78, 79, 101, 106
local government, viii, x, 2, 3, 14, 76, 77, 79, 109, 111
location, 7, 66

losses, 3, 6, 72

M

magnetic, iv
maintenance, vii, viii, ix, 1, 2, 3, 4, 5, 7, 10, 12, 13, 23, 24, 25, 26, 27, 28, 29, 32, 33, 38, 39, 41, 49, 54, 55, 60, 66, 67, 68, 69, 73, 118, 119, 120, 121, 122, 124
management, ix, 10, 23, 24, 26, 34, 45, 46, 52, 56, 59, 63, 65, 67, 68, 80, 81, 88, 91, 93, 95, 96, 100, 101, 102, 114, 115, 118, 119, 120, 121, 122, 123, 124, 125, 126
management practices, 26
mandates, 87
market, 16
marshes, 63
measures, 31, 100, 108
media, 80
membership, 59, 60, 63
memory, 127
Mexican, 90, 91, 92, 100
mine tailings, 19
mining, 3, 12, 114
missions, 35, 118, 120
modeling, 125
models, 27
modernization, 60, 61, 62, 119
money, 9, 47, 54, 55, 56, 77, 78, 92, 93, 94, 95
morning, 65, 71
movement, 10
MSC, 35
multiple factors, 124

N

NAFTA, 92
nation, viii, 1, 2, 3, 5, 6, 7, 8, 10, 14, 18, 24, 28, 31, 33, 38, 52, 77, 79
National Academy of Sciences, 117
National Park Service, 11
national policy, 61
National Research Council, 35, 117
National Wildlife Refuge, 39, 40, 44, 49
Native American, 62
natural, 18, 40, 41, 44, 50, 59, 72, 87
natural disasters, 18
natural gas, 87
Natural Resources Conservation Service, ix, 4, 19, 24, 30, 32
navy, 11
non-defense, 106
non-profit, 15
normal, 13, 14, 28, 33, 43, 99, 120
Nuclear Regulatory Commission (NRC), 4, 8, 122
nuts, 24

O

obligations, 28, 33, 54, 55
observations, 123
Office of Management and Budget, 31, 38, 60, 61, 99, 114
oil, 54, 67
omnibus, 86, 93, 94, 106, 108
online, 19
openness, 120
operator, 30, 38, 47
outsourcing, 124
oversight, 68, 124, 126, 127
ownership, 3, 13, 16, 19, 56, 73

P

partnership, 39, 53, 56, 57, 93, 95, 96, 99
passive, 38
peer, 122
peer review, 122
Pentagon, 104
performance-based budgeting, 31
periodic, 7, 8, 10, 26, 122
permit, 12, 27

philosophy, 68
planning, 15, 25, 30, 52, 53, 56, 69, 120, 121, 125
planning decisions, 25, 69
plants, 24, 68, 77, 78, 118, 119
play, 54, 123
policy makers, 117
pollution, 77, 81, 100, 101, 102, 114
population, 53, 119
population growth, 53
portfolio, ix, 23, 24, 26, 31, 34, 120
potatoes, 40
power, 2, 24, 31, 53, 54, 55, 59, 60, 62, 65, 67, 68, 69, 70, 119, 120, 121, 123
power plant, 24, 119
power stations, 31
precedents, 27
pressure, 38, 68, 78, 83
prevention, 4, 5, 62
preventive, 68
prices, 54
private, 3, 4, 5, 12, 14, 15, 16, 19, 26, 43, 48, 56, 99, 118, 125
private citizens, 48
private ownership, 16
private sector, 118, 125
production, 43, 63
profit, 15
property, iv, 3, 9, 19, 29, 67, 72
prosperity, 53
protection, 3, 52, 55, 62, 67, 68, 84, 93, 98, 114, 118, 123
public, vii, 1, 2, 3, 4, 5, 14, 18, 24, 26, 27, 29, 32, 39, 43, 44, 46, 49, 52, 53, 54, 56, 61, 67, 77, 78, 79, 91, 114, 117, 118, 119, 120, 123
public health, 56, 78, 79
public investment, 53
public policy, 27
public safety, vii, 1, 2, 5, 18, 26, 32, 52
public-private partnerships, 56
pumping, 118, 119

Q

quality improvement, 82
quality of life, 52, 53
query, 9

R

range, 118
ratings, 5, 18
Reagan Administration, 77
real estate, 15
reality, 61
recognition, 12, 40, 63
recovery, 5, 45, 72, 73, 122
recreation, 2, 27, 43, 53, 63
recreational, 3, 39, 40, 44, 46, 49
reflection, 59
refuge, 44
regional, 2, 5, 12, 26, 52, 61, 66, 120, 122, 124, 126
regional economies, 2, 5
regular, vii, viii, 1, 2, 25, 68, 73, 122
regulations, 12, 61, 77, 78, 79, 119
regulators, 11
regulatory commission, viii, 2, 4, 8
regulatory requirements, 119
rehabilitate, viii, 2, 6, 14, 17, 18, 25, 38, 39, 46, 47, 49, 60, 69
rehabilitation, viii, 1, 2, 6, 9, 10, 13, 14, 17, 18, 25, 26, 27, 28, 29, 30, 32, 33, 38, 39, 45, 46, 47, 48, 49, 53, 55, 60, 61, 62, 68, 69, 119
Rehabilitation Act, 12, 28
rehabilitation program, 17, 61, 69
reimbursement, 14, 29, 33
relationships, 54, 118, 121
reliability, 54, 68, 70, 122
repair, 6, 8, 14, 17, 18, 25, 28, 29, 33, 38, 41, 42, 45, 66, 68, 69, 72
rescission, 102, 109, 110
reservoirs, 24, 27, 40, 44, 53, 67, 68

residential, 43, 65, 78
resolution, 48, 86, 111
resources, vii, viii, ix, 2, 4, 12, 23, 24, 25, 30, 31, 32, 33, 34, 39, 43, 49, 51, 52, 53, 54, 55, 56, 59, 60, 67, 69, 71, 87, 93, 119, 121, 123
responsibilities, viii, ix, 2, 7, 11, 23, 26, 30, 56, 69, 77, 119, 126
restructuring, 61
retention, 17, 124
revenue, 16, 55
rings, 119
risk, vii, 1, 2, 13, 14, 18, 28, 32, 45, 46, 56, 66, 69, 126
risk assessment, 13, 32, 56
roadmap, 121
rodent, 66, 72
royalties, 54
rural, 52, 56, 62, 63, 77, 82, 100
rural areas, 56
rural communities, 52, 82

S

Safe Drinking Water Act (SDWA), vii, x, 76, 78, 79, 80, 81, 82, 91, 93, 95, 96, 97, 98
safety, vii, viii, 1, 2, 3, 4, 5, 6, 7, 8, 9, 10, 12, 13, 14, 15, 17, 18, 19, 20, 26, 28, 31, 32, 52, 54, 67, 68, 72, 73, 118, 119, 120
sales, 16, 54
savings, 56
SBR, 60
security, 4, 5, 9, 26, 98, 118, 120
seed, 14, 28, 29, 40, 47, 78, 82
seismic, vii, 1, 5, 10, 14, 28
seismic data, 14, 28
semiarid, 62
senate, v, vi, ix, x, 21, 34, 35, 37, 48, 51, 52, 59, 65, 71, 73, 83, 84, 86, 94, 96, 97, 98, 99, 100, 101, 102, 103, 104, 105, 106, 107, 108, 109, 111, 112, 114, 129
series, 29, 42, 52, 94, 107
services, iv, 47, 56, 68, 69, 84, 114, 125, 127
settlements, 54, 62
settlers, 41, 62
sewage, 77, 78, 87, 88, 90
shape, 35, 92
shares, 83
sharing, 17, 30
short-term, 94, 113
signs, 10
siphon, 42
sites, 4, 5, 114
skills, 124, 127
social benefits, 2
solar, 115
spawning, 45
species, 45, 53, 68, 122
SRFs, x, 76, 82, 85, 91, 98, 101, 109
stability, 8, 42
staffing, 125
stakeholder groups, 101
stakeholders, 28, 38, 47, 52, 56, 70, 101, 118, 121, 124, 125, 126
standards, 6, 7, 13, 20, 25, 56, 78, 121, 123
state department, 4
state grants, 15
state laws, 67
state regulators, 11
statutory, 89
steel, 29, 42
steel pipe, 29
stimulus, 90
storage, 2, 3, 24, 44, 53, 67
stormwater, 16
strength, 59, 69
summer, 27
supplemental, 27, 40, 41, 43, 44, 86, 90
supply, 2, 3, 15, 16, 31, 38, 41, 46, 52, 56, 57, 60, 62, 72
supreme court, 99
surface water, 63
systems, 26, 63, 68, 77, 78, 79, 103, 119, 120, 125

T

targets, 87
tax-exempt, 55
technology, 9, 68
Tennessee Valley Authority, viii, 2, 4, 8, 11, 118
territory, 11
terrorism, 5
terrorist attack, 104
testimony, ix, x, 35, 37, 52, 59, 118, 129, 130
threat, 3, 39, 67, 77
threatened, 43, 45, 53
threshold, 6
time, x, 10, 13, 32, 39, 45, 48, 49, 50, 56, 60, 68, 72, 73, 76, 78, 81, 82, 83, 85, 94, 102, 112, 118, 122
timing, 69
title, 26, 83, 89
TMC, 65, 66
tourism, 44
tracking, 120
training, 7, 9, 10, 17, 18, 115, 126, 127
transfer, 33
transition, 77, 119
transport, 38, 40
transportation, 29, 63
travel, 48
treasury, 54
treaties, 54
tribal, 30, 54, 93, 95, 97, 100
tribes, 93, 94
trout, 43
TSC, 124, 125
turnover, 124

U

U.S. Department of Agriculture, 30
U.S. Treasury, 54
uncertainty, 53
uniform, 7, 121
urban areas, 114
urbanized, 67

V

values, 53
vegetables, 24
vice president, 59, 63, 129
vision, 40
vulnerability, 10

W

wastewater, vii, ix, x, 16, 52, 75, 76, 78, 79, 81, 82, 83, 84, 87, 88, 89, 91, 92, 93, 96, 98, 100, 101, 102, 103, 109, 111, 114, 115
wastewater treatment, vii, ix, x, 75, 76, 78, 79, 81, 83, 84, 87, 88, 89, 109, 111, 115
water quality, x, 76, 78, 80, 81, 82, 86, 88, 112, 114
water resources, vii, viii, ix, 23, 24, 25, 30, 31, 32, 33, 39, 49, 53, 55, 56, 59, 71, 119
water rights, 29, 30, 38, 39, 40, 44, 47, 49, 62
water supplies, 24, 31, 53, 63, 79
waterfowl, 44
watershed, 4, 35, 41, 52, 56
wealth, 39, 49
wear, 119
web, 34
webpages, 19
welfare, 119
white house, 86
wholesale, 67
wildlife, 3, 43, 53, 63
winter, 43
workforce, 119, 124
workload, 118, 119, 120, 123, 124
World Trade Center, 104
worry, 72